INVENTEERING

A Problem-Solving Approach to Teaching Technology

BOB CORNEY
NORM DALE

Trifolium Books Inc.
Toronto, Canada

Trifolium Books Inc.
250 Merton Street, Suite 203
Toronto, Ontario, Canada, M4S 1B1
Tel: 416-483-7211 Fax: 416-483-3533
E-mail: trifoliu@ican.net www.trifoliumbooks.com

Special Note: This resource has been reviewed for bias and stereotyping.

Canadian Cataloguing in Publication Data
Corney, Bob, 1934–
Inventeering: a problem-solving approach to teaching technology
(Springboards for teaching)
ISBN 1-55244-014-1

1. Technology—Study and teaching (Elementary). I. Dale, Norm. II. Title. III Series

T65.C676 2000 372.3'58044 C00-930341-3

Project Editor: Julie E. Czerneda
Design, layout, graphics: Heidy Lawrance Associates
Project coordinator: Jim Rogerson
Production coordinator: Heidy Lawrance Associates
Cover design: Fizzz Design

Trifolium Books Inc. acknowledges the financial support of the Government of Canada through the Book Publishing Industry Development Program (BPIDP) for our publishing activities.

Canadä

Printed and bound in Canada
10 9 8 7 6 5 4 3 2 1

Trifolium's books may be purchased in bulk for educational, business, or promotional use. For information, please write: Special Sales, Trifolium Books Inc., 250 Merton Street, Suite 203, Toronto, Ontario M4S 1B1

Safety: The activities in this book are safe when carried out in an organized, structured setting. Please ensure you provide your students with specific information about the safety routines used in your school. It is, of course, critical to assess your students' level of responsibility in determining which materials and tools to allow them to use.

Note: If you are not completely familiar with the safety requirements for the use of specialized equipment, please consult with the appropriate specialty teacher(s) before allowing use by students. As well, please make sure that your students know where all the safety equipment is located, and how to use it. The publisher and authors can accept no responsibility for any damage caused or sustained by use or misuse of ideas or materials mentioned in this book.

Acknowledgment: The publisher and the authors would like to express their appreciation to Dan Forbes of Winnipeg, Manitoba, for his careful and insightful review of this project.

Table of Contents

Meet the Authors 5
Introduction 6
The Linking of Science and Technology 8
Direction and Focus in Technology: How Does Technology Fit In? 9

Section 1: The Design Process 11
- Exploration Skills 11
- The Design Process 11
- The N. I. C. E. Model 12
- The Teacher's Role in the Design Process 14
- An Application of the Design Process 15
- A Teacher-Directed Design Process 17

Section 2: A Safe and Organized Classroom 18
- Safety 18
  - Goggles/Tool Handling/Housekeeping/Deportment
- Classroom Organization 19
  - The Travelling Workstation/Independent Workstation
  - Storage Areas/Work Surfaces 21
- Metric Measurement 22

Section 3: Basic Tools of Technology 23
- Junior Hacksaw/Dovetail Saw/Bench Hook/
- Hand Drill/Drill Bits/Corner Joiner/Gussets/Hole Punch
- Metal Safety Ruler/C-clamp

Section 4: Specialized Tools of Technology 26
- Table Vice/Glue Gun/Hammer/Nails/Screwdriver
- Wood Screws/Needle-nose Pliers/Side-cutting Pliers
- Wire Strippers/Snips/Utility Knife/Files

Section 5: Materials 29
Junk Materials 29
Manufactured Materials 30
Wood Strips/Dowel Rods/Gussets/Art Straws/Wheels/Pulleys/Gears/Electrical Components and Supplies/Syringes and Tubing

Section 6: Tools and Materials Applications 34
Basic Tool Techniques 34
Specialized Tool Techniques 38

Section 7: Building Structures from Different Materials 39
Cardboard Materials 41
Art Straws 42
Wood Strips 42

Section 8: Machines and Mechanisms 44
Levers/Wheels and Axles/Pulleys/Gears/Cams Winches/Hinges

Section 9: Energy and Control Systems 57
Mechanical Energy 57
Elastic Bands/Syringes and Tubing
Electrical Energy 59
Component Voltages/Circuit Diagrams/Electrical Symbols/Simple Switches/Series and Parallel Circuits/Fastening Electrical Components to Structures/Control from a Remote Location/Reversing the Direction of Motor Rotation

Section 10: Approaches to Assessment 65
Checklists 66
Rating Scales (Rubrics) 70
Summary 75

Section 11: Activities & Challenges 76
Teacher Directed Activities by Strand/Topic/Grade Level 77
Summary of Activities 78
Open-ended Problem-solving Challenges 100
Summary of Challenges 101

Section 12: Appendices 115
Useful Forms and Diagrams 115
Additional Resources 122

MEET THE AUTHORS

Bob Corney

After 17 years in Wood-Patternmaking with The Steel Company of Canada in Hamilton, Bob obtained his teaching certificate from the University of Toronto and began teaching Cabinetmaking/ Building Construction with the Wellington Board of Education. During this time, he obtained his Specialist Certification from the Faculty of Education at Queens University. The following and remaining years until retirement, Bob continued in Technogical Education as a teacher, Technical Director, Consultant, and Coordinator for the Peel Board of Education.

Currently, Bob is a part time instructor for OISE/UT, teaching in the Initial Certification Program for Technological Studies candidates, as well as a component of the Honours Technological Studies Specialist Qualification. He is also a partner with co-author Norm Dale in CorDale Enterprises, an educational consultancy with a focus on assisting elementary school teachers with the implementation of technology applications.

Norm Dale

Norm was born and raised in west-end Toronto, where he began his apprenticeship as a Construction Electrician and remained in the trade for ten years before entering the Faculty of Education at the University of Toronto. He taught Electrical Construction and Maintenance for 14 years. A one-year stint as President of District 14 O.S.S.T.F. was followed by 5 years at Frank Oke Vocational School as Chairman of Technological Studies. Norm went on to become the Consultant for Technological Education for the York Board, a position he retained until his retirement. During this time, Norm was involved in the creation of the Ontario 'Technological Education Coordinators' Council and served a two-year term as President.

Norm maintains his involvement with the application of technology in elementary schools, as directed by the Science and Technology Grades 1 to 8 curriculum guidelines, through his partnership with co-author Bob Corney in CorDale Enterprises.

INTRODUCTION

A new focus for curriculum direction in Elementary schools has been the linking of Science with Technology. This direction identifies several "expectations" for teachers to deliver unfamiliar processes in Technology programs across these grade levels. Very likely, most teachers will have no difficulty adapting Science concepts to these identified expectations. Technology, however, has never been clearly identified as a curriculum focus in elementary education, with the result that General-Studies teachers are being asked to satisfy expectations for which they have little or no training. Many of the Technology concepts, skills, and strategies within the curriculum focus will be ignored, if not lost, unless ways are found to assist teachers with this focus. How teachers are to gain a comfort level with the technologies, so that our students are well prepared to deal with the problems of human adaptation to new technology in society and the environment, is an upcoming challenge. This resource is dedicated to that need.

What is Technology?

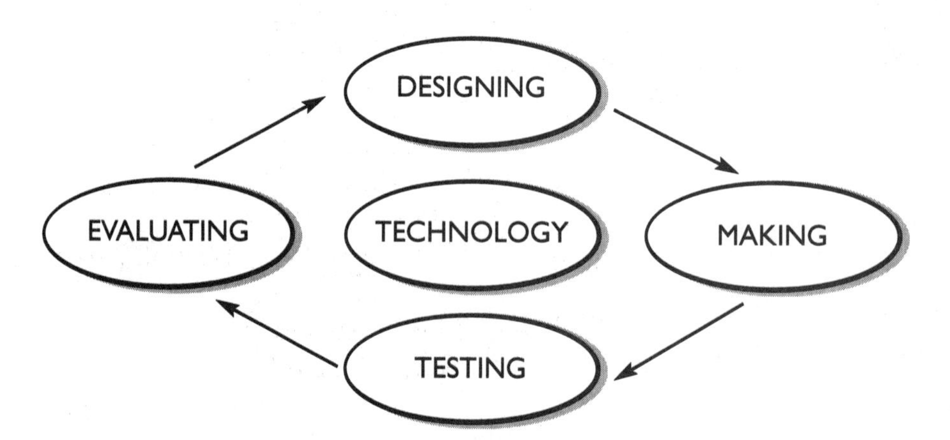

Technology, in this context, is about designing, making, testing, and evaluating. Students are being asked to apply knowledge obtained from Science (and other disciplines) to solve a specific problem or challenge using materials and tools (including computers). Students will be challenged to suggest ideas and to go beyond these to seek alternatives. As they begin to develop a technological awareness, students will also begin to question processes, their actions, and the impact of their decisions on society and the environment. The elementary school years are those in which teachers can assist students with their technological decisions and actions by responding to and influencing those decisions and actions. This is where most teachers will need direction and focus, and that is the intent of this resource.

A resource such as *Inventeering* has often been requested because of the uncertainty many teachers have expressed in having to meet curriculum expectations they don't feel equipped to handle. What's not clearly evident in the curriculum is how the technological experiences are to be provided; what technological experiences should be provided; what methods and procedures should be used to provide these experiences; and how we assess these experiences.

Thinking about ...

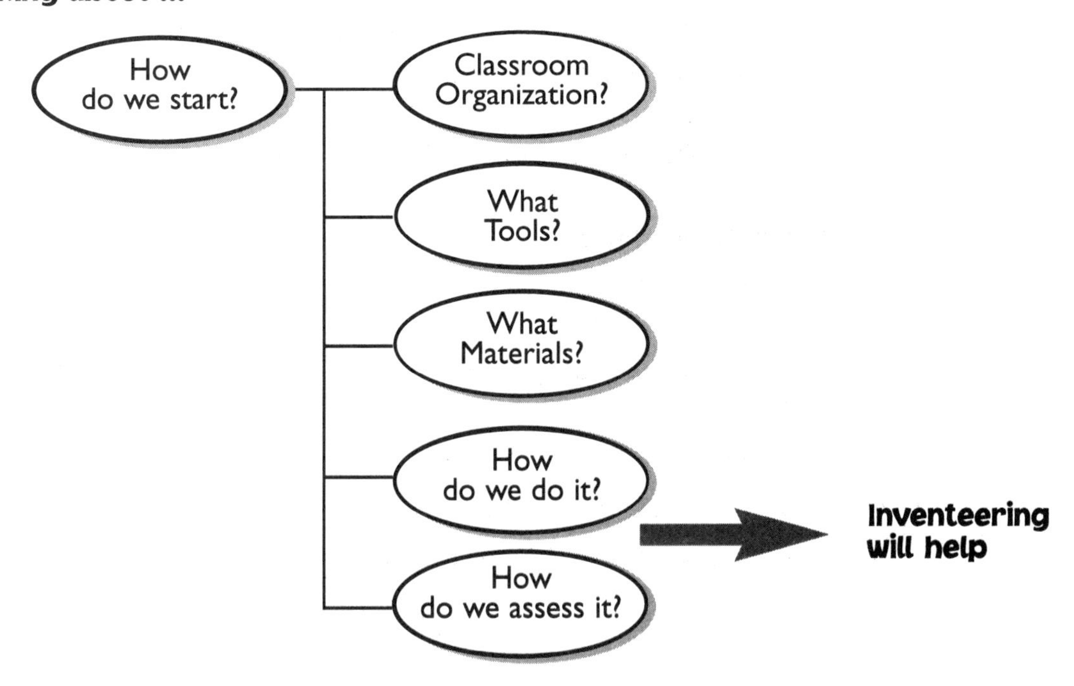

Thus, *Inventeering* was written with a focus on "getting started" with technology. In this resource you will find suggestions for classroom safety and organization, for the tool sets you will need, and for materials selection, all with a focus on the Design Process and its open-ended problem-solving explorations. Several suggestions for activities and challenges linked to student expectations will help in getting started. Assessment and evaluation techniques are guided by curriculum directions and are similar to processes teachers are likely already using, but in some ways unique to technology. Some authentic evaluation methods are included for consideration.

THE LINKING OF SCIENCE AND TECHNOLOGY

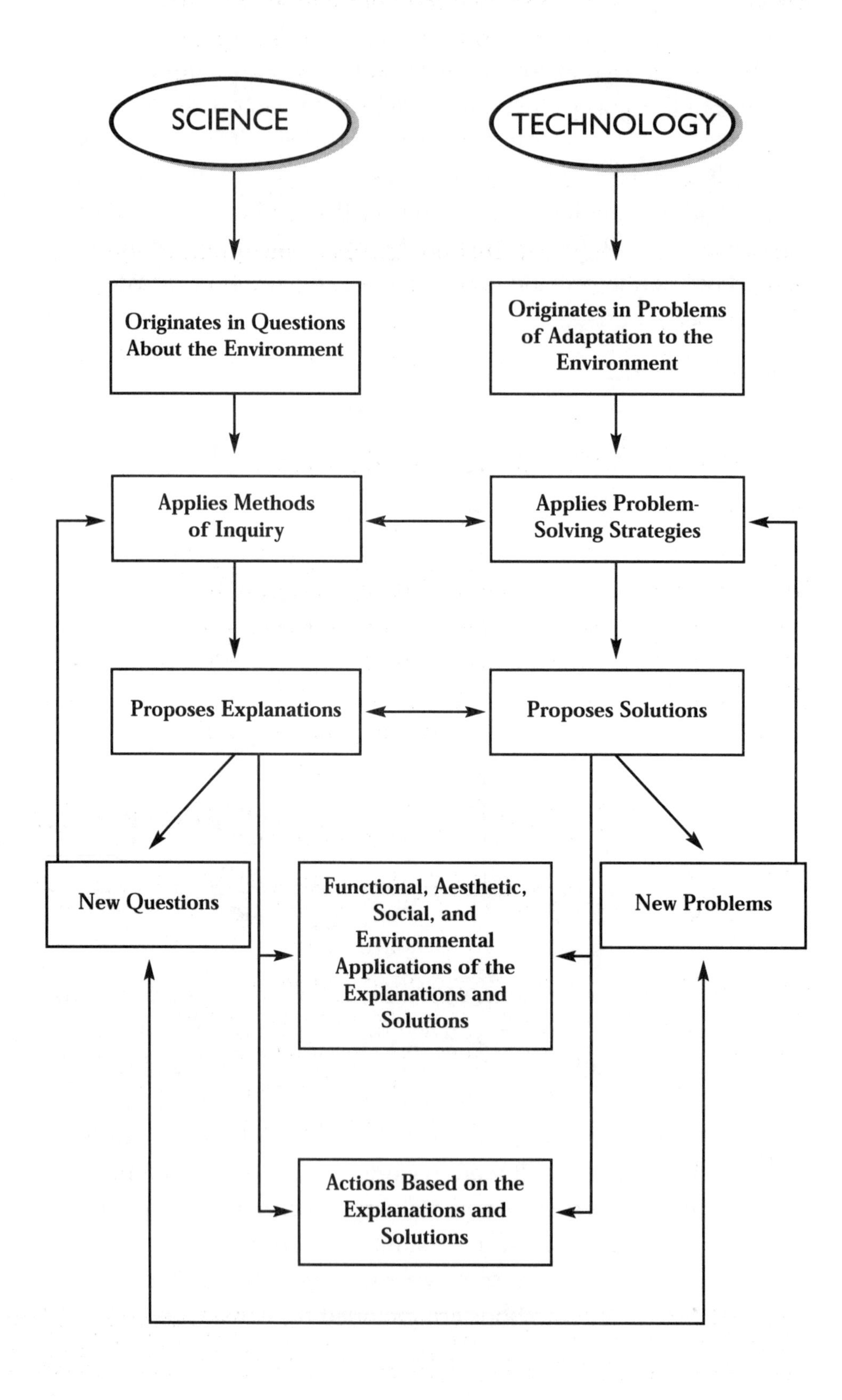

The foci of Science and Technology, as shown, highlight the natural links between both disciplines. In Science, the questions asked, the inquiry method used to attempt to answer these questions, and the explanations proposed, are seeking to explain the "why" of relationships and patterns in the environment. Technology, in adapting to these relationships and patterns, produces solutions to practical problems and through a process of problem-solving, answers the question, "how." Science has expanded our understanding of the environment and the way in which things interact within it. Similarly, Technology has made significant changes in our lives through the discovery of new applications. It is in this context that students must see Science and Technology as interrelated both within and beyond the school environment.

DIRECTION AND FOCUS IN TECHNOLOGY

HOW DOES TECHNOLOGY FIT IN?

A question often asked by teachers of elementary programs is "how am I going to fit this technology into an already cluttered curriculum?" This is a valid concern and will only add to the anxiety unless the message is given that the technology component is not "as well as;" rather, it is "along with" other curricula and should be considered as an enhancement and support rather than a replacement of current curriculum directions.

Similarly, teachers must understand that an application of technology is not a process described by a commitment to a time frame. As students experience a search for knowledge or solutions to a problem, their application of Science, of Mathematics, of the Social Sciences, of Reading and Writing, along with the technology, is of equal if not a greater value than an isolated practice of technical skills. With technology applications, the acquisition of academic skills is still taking place but not always as a single focus as is often the case in traditional settings. This multidisciplinary and interdisciplinary approach cements the links between the curriculum pieces.

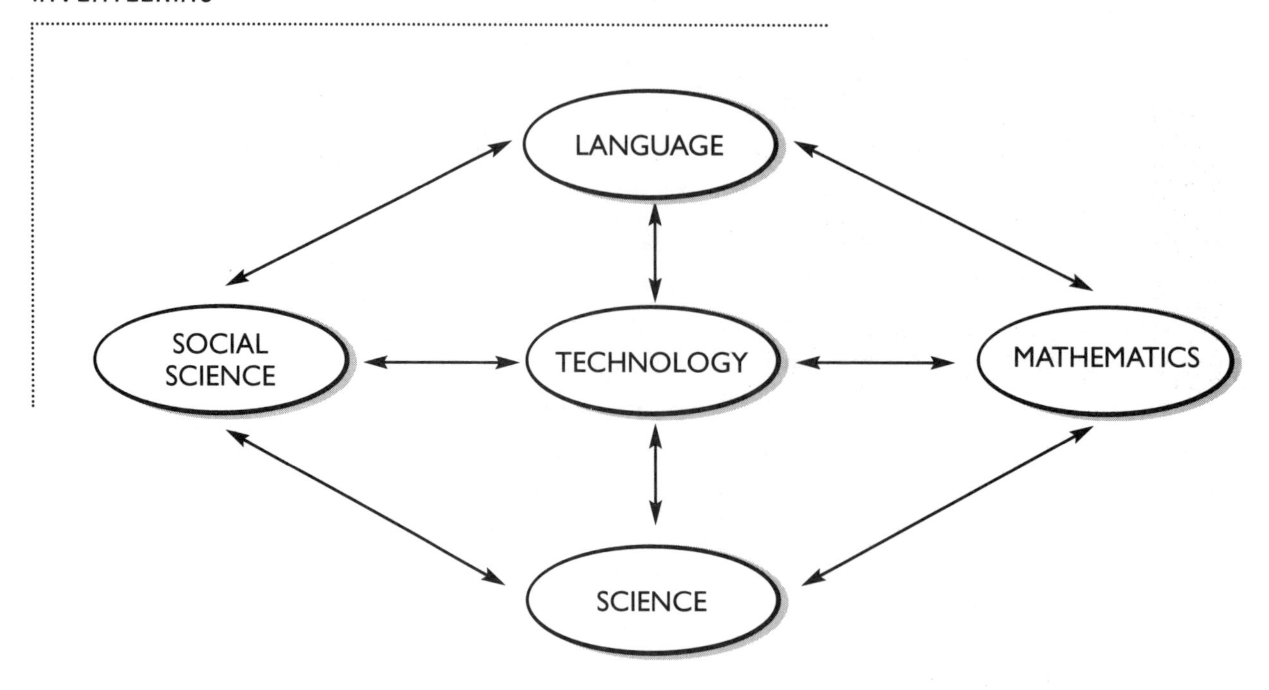

Technology applications should arise from work already taking place in the classroom. For teachers to be comfortable with this new direction, they need to be aware of this potential within existing curriculum. The existence of a stimulus or trigger for technological activities will provide opportunities for students to acquire and practice a range of practical skills. If students were studying Life Systems in Science, or more specifically the different ways that animals move to meet their needs, the technology stimulus might be to make an object or device that simulates their movement (a frog hopping, an insect flying, etc.).

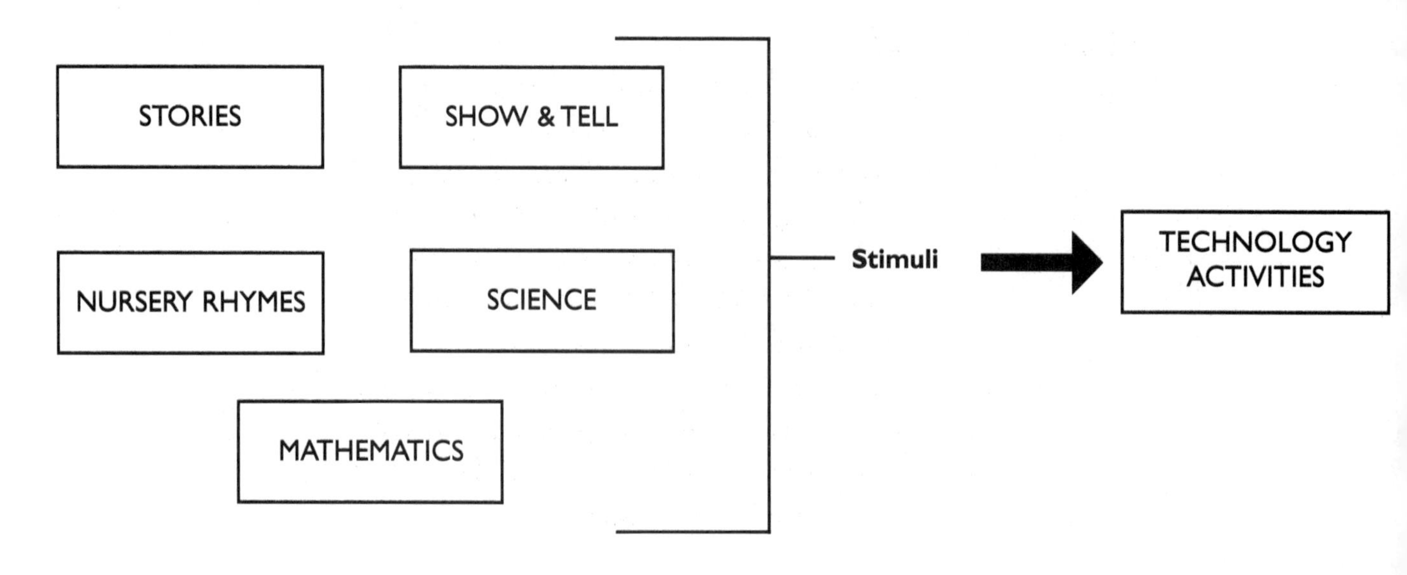

Section 1

THE DESIGN PROCESS

EXPLORATION SKILLS

Inquiry and Design are key components of the Science and Technology processes. In both cases, problems are identified, explored, and refined into component parts small enough to enable planning the best strategy for researching and solving them. In both inquiry and design, the processes involved provide students with opportunities to develop much needed exploration skills. In technology, the basis for developing solutions to practical problems is the Design Process. This is the fundamental building block of Technology.

THE DESIGN PROCESS

The Design Process is often considered a "journey" with a number of stops along the way to help students, even at an early age, to be creative when there is a problem to be solved. There are several Design Process or Problem-Solving models available for teachers to consider. Each differs only in their complexity and often, number of stops in the journey. A good Problem-Solving model is measured by its ability to satisfy the technological experience intended for students. For instance, some models have component parts that are beyond the capabilities of students in the early grades and hence are not suitable. The following describes the **N. I. C. E.** Design Process. This is one of the simplest Problem-Solving models with which to begin.

THE N.I.C.E. MODEL

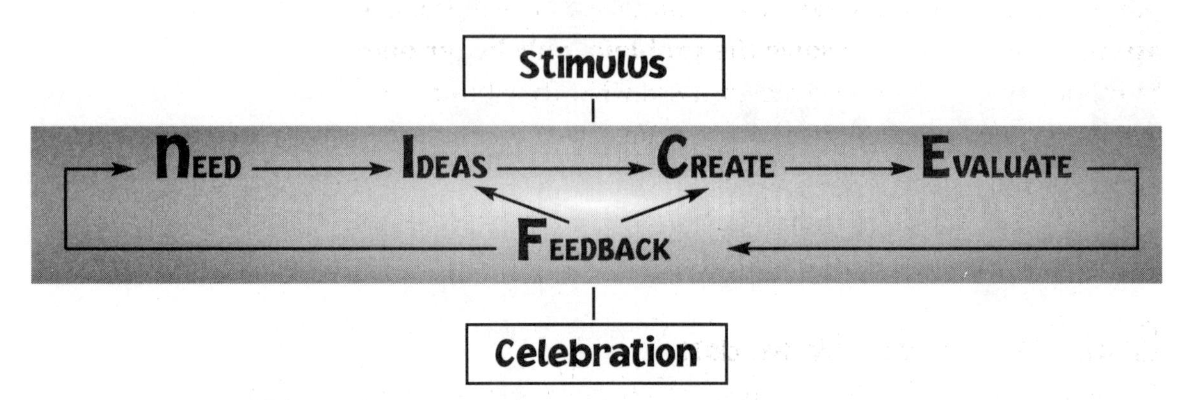

NEED: (What do we have to do?)

A problem to be solved always begins with a situation that identifies a "NEED". This "need" often grows out of a stimulus identified from a current curriculum focus and sets the stage for the technological problem-solving challenge that follows. The identified "need" must be relatively simple for younger students and set within the context and relevance of the curriculum and the capabilities of the age group. The "need" will increase in complexity as students mature and their capabilities develop.

IDEAS: (How will we do it?)

Many "IDEAS" on how to solve a problem flow out of the identified "need". **In technology, we need to remember there is no single solution to a problem to be solved and we should encourage students to think about many possible alternatives while discouraging the natural tendency to develop only their first idea.** The "ideas" stage is where "designing" takes place and requires students to gather and research as much information as possible from resources both within and outside the school. The many ideas posed in their design ideas will need decisions to narrow their solution choices. During this stage, students will make preliminary sketches that are continually modified as new ideas are explored. The final sketch or drawing is the blueprint for the "create" stage of the design process. Consideration must also be given to the availability of tools and materials.

CREATE: (How can we make it?)

Once a solution has been chosen that students think will best solve the problem, they now have to build it. This is the "CREATE" stage and allows for further decision making in choice of materials and assembly procedures. **In the create stage, we need to remind ourselves that there are no wrong solutions, or materials, or assembly procedures to solve the problem, only better ones.** Students must be encouraged to value what they have done, learn to be their own critics without feeling discouraged, and to always see the potential for doing better next time. Once students evaluate their product, they themselves will come to see how it might have been approached differently.

EVALUATE: (How well did we do?)

At the end of the Design Process, the chosen solution to the problem has been made. Students will now want to decide how successful their chosen solution was in satisfying the original "need". Seldom is a solution perfect and often modifications can be made to improve it.

FEEDBACK

Sometimes it may be necessary at each of the stages, to "back track" to previous stages to make modifications to earlier decisions that might improve the results.

CELEBRATION

The design journey is complete when students demonstrate their solution to their teacher and classmates and celebrate their successes.

It has long been observed that students' ability to recognize, analyze, and solve problems is an essential component of their intellectual development. The planning, development, and evaluation of a practical solution to a challenge, are distinct stages within the Design Process.

The Design Process incorporates a method of open-ended problem-solving and the collaborative nature of this process, provides students with an ideal opportunity to ask questions, plan investigations, use appropriate vocabulary in describing their investigations, and to communicate the procedures and results of their investigations to their peers. The development of valuable thinking and social skills, is one of the products of this process.

Within the Design Process, emphasis should be placed on sketching or drawing as a means of developing recording and writing skills. Students who have difficulty with expressive writing often have no difficulty in providing a two-dimensional sketch or scale drawing of what they are going to do or what they have done. The sketch or drawing, along with appropriate labeling, can convey a great deal of information and also links to many curriculum expectations. For instance, the sketch or drawing will provide opportunities for students to measure, classify, identify, compare, and determine through observations, the interactions of their material choices and the assembly procedures for their chosen devices. The development of a vocabulary of new terms also go hand in hand with this process.

The evaluation stage of the Design Process allows for the development of the student's spatial awareness. They will be taking a multi-sided view of the challenge and interpreting how each component part fits within the total solution. In addition, students enjoy making verbal and visual responses in describing both their learning process and the product of their efforts, thus fulfilling many of the "communicate" expectations within the curriculum.

THE TEACHER'S ROLE IN THE DESIGN PROCESS

The Design Process has to be taught for students to understand the benefits of the process. Role-playing through a simple problem--solving exercise is a good way for students to see what their roles and expectations might be at each step of the process. They need to be encouraged to listen to and to value the ideas offered by each group member. They have to be shown that there are several possible solutions and should be discouraged from proceeding with only their first idea. Acceptance of the majority decisions of the group, recognition of the abilities of other group members for an equitable division of duties, and learning that it's normal to rethink the direction they have taken if things are not going as expected, are natural results of involvement in the process and not an indication of failure.

Once students have learned and begun to apply the Design Process, the teacher has to be a "guide on the side" and become a facilitator rather than a director of the process. There is a natural tendency for students to ask "how can I do this" and "what should I do next." The teacher has to suggest "I don't know the answer to your question, but lets see if we can work this out together." With this, the teacher has placed him or herself in their world as a learner and not an expert. If the process has bogged down, hints or suggestions for them to try is often all it takes for them to get back on track.

An Application of the Design Process

SITUATION: You are currently studying Life Systems with your students. Coincidentally, some of your students have returned to school talking about seeing some frogs in the park and how these frogs were able to avoid being captured. An idea that fits within the "Characteristics and Needs of Living Things" component of the Life Systems unit is to challenge your students to make a device that simulates a hopping frog. This stimulus or trigger for a technological activity has provided an opportunity for your students to acquire and practice a range of practical skills while also using this activity to connect to the knowledge component of the curriculum focus. (i.e. ways in which animals move)

Need

To build a device that simulates a frog that hops.

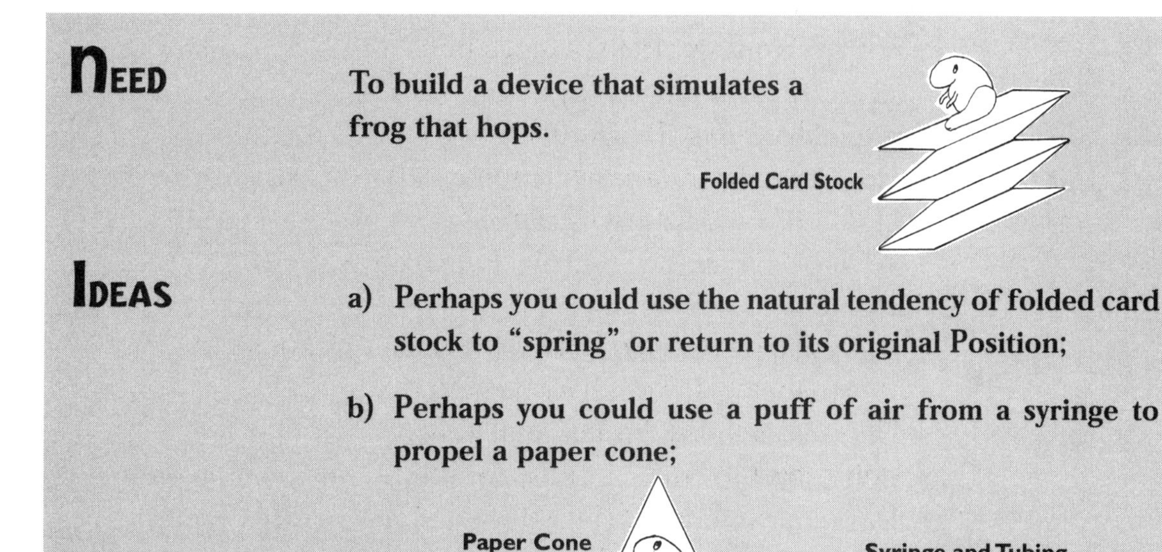

Ideas

a) Perhaps you could use the natural tendency of folded card stock to "spring" or return to its original Position;

b) Perhaps you could use a puff of air from a syringe to propel a paper cone;

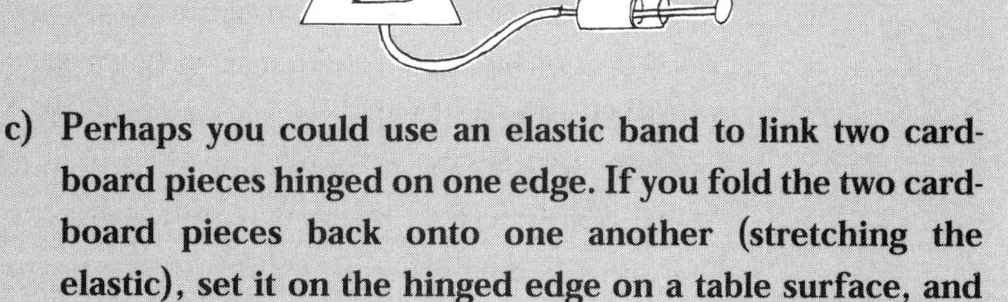

c) Perhaps you could use an elastic band to link two cardboard pieces hinged on one edge. If you fold the two cardboard pieces back onto one another (stretching the elastic), set it on the hinged edge on a table surface, and then let it go, it will hop.

Sketches for each possible solution should be encouraged.

CREATE

Assuming a choice has been made in each collaborative group as to which idea best provides a solution to the need, decisions will have to be made for materials selection and assembly procedures.

EVALUATE

Students evaluate their product by considering: Were the materials chosen appropriate? Did the solution satisfy the need (hop like a frog)? Could the solution be made better? How does the solution compare with those of other designs?

Celebration

Time should be set aside for students to share their experiences with their classmates. The teacher can be a passive observer or can direct the presentation by asking questions of the group. These questions might include:

- How did you decide what your solution to the challenge would be?
- How did you decide which materials were to be used?
- What problems did you have in making it work?
- How did you overcome these problems?
- Did you need to change or modify your solution?
- What modifications did you make?
- Did your solution work as well as expected?
- If you were to make it again, would you do anything different?
- Does your solution meet the need as well as other group's solutions?
- What did you learn from this experience?

Throughout the sharing process, student groups should be encouraged to display and refer to any preliminary and final sketches or drawings of their solution and how it satisfied the original need. Each should also be prepared to answer questions invited from their classmates. Sharing becomes important in developing the skills of listening, questioning, analyzing, communicating, co-operating, as well as building a vocabulary of terms.

A TEACHER-DIRECTED DESIGN PROCESS

Although technological competencies develop as a student progresses through their schooling, it is the student in the early grades or the untrained beginning student who is easily discouraged and needs a lot of teacher help. These students are limited in their abilities to independently identify needs or problems and to come up with practical solutions. They are very good at observing what is going on and in asking questions, but are limited in developing design ideas because of their stage of development, knowledge, and experience. Lacking in all but the simplest manual skills, they will try producing simple products by reproducing model examples or through step-by-step supervision.

The task a teacher faces in helping these young students with limited technological capabilities, is to increase their skills in small progressive stages. To begin the transformation, the problems to be solved, the applications of the design process, and the possible solutions, should be teacher guided in order to make students comfortable with solving simple problems, making simple decisions, and offering ideas. This, in effect, is a teacher-directed design process.

Section 2

A SAFE AND ORGANIZED CLASSROOM

SAFETY FIRST!
Check these boxes for tips and specific warnings related to classroom safety issues.

SAFETY

Safety should be taught to students and not learned the hard way. Many school boards have developed policy manuals for the safe operation of classroom and specialized facilities. If available, these should be reviewed for the safety precautions appropriate to the technological activities taking place in your classroom. The following is an overview of the more common safety rules to be considered.

GOGGLES

Normally, goggles do not have to be worn at all times, but certainly should be worn whenever there is a danger of material coming into contact with the eyes. Teachers need to develop hard and fast rules for goggle use in the early grades and their use insisted upon when appropriate. Some workstation activities would suggest their use. Although using hammers for many of the activities outlined in this resource is minimal, it could occur and eyes need protection. It is also wise to wear goggles when glue guns are being used. Sawing wood is not normally a situation that requires the use of eye protection. There have been occasions though, when sawdust blown from the pieces being cut has flown into eyes. Common sense is your best guide to when goggles should be used. Generally, it is better to be safe than sorry.

TOOL HANDLING

Students should be taught the safe and proper use and handling of all tools. It should not be assumed that students know their proper use. Each tool is designed for a specific function and should not be used for other purposes. For instance, utility knives should not be used as levers (prying lids off cans) when a screwdriver is better suited. Pliers should not be used as hammers. Saws should only be used for cutting wood. When walking from place to place, students should carry tools at their sides to avoid injury to other classmates.

HOUSEKEEPING

Maintaining a tidy and safe work area should be a priority. Tools and materials that are not in use should not be left on work surfaces to create clutter or to be dropped on the floor and tripped over. Putting tools back in their stored location, when not needed, promotes organization skills and a reduction in the loss of time looking for them.

DEPORTMENT

Many accidents happen when boisterous student behavior around tools and materials is not checked immediately. There is a tendency for students to run from place to place in their enthusiasm for building and testing their devices. Tools and materials have been thrown rather than placed on work surfaces. Teachers need to set and reinforce safety expectations.

CLASSROOM ORGANIZATION

Whether your school has a dedicated room as a Technology Resource Centre or a regular classroom, the following are suggestions to help you visualize what you might need to provide for your technology applications. A well-organized classroom is also a safer one, in which work areas and tool storage are accessible and clearly identified.

Classroom organization will depend on the space available, the resources available, and whether or not you share the resources with other teachers.

THE TRAVELLING WORKSTATION

If you are required to share resources between classrooms in your school, a "workstation on wheels" is available from several educational suppliers. These range from wire-frame carts with plastic storage bins, to wooden cabinets designed to provide a sturdy work surface with lockable doors and drawers to house all the required tools and materials. Although more expensive, the cart with a work surface is ideal, particularly if classroom space limits the provision of separate work surfaces. The use of the travelling workstation requires someone in the school to schedule its use as well as to replace the materials as they are used.

SAFETY FIRST!

If you have a travelling workstation, ensure a teacher or other adult is in charge of moving and storing it. Do not allow students to move the workstation on their own.

INDEPENDENT WORKSTATIONS

Where space permits, organizing your classroom with a series of separate workstations not only provides student work areas, but also allows you to identify and equip several locations with the tools and resources required for specific technology activities. For instance, you may have a Construction Centre where the assembly of solutions to a challenge takes place; an Energy and Control centre for students to make applications with electrical and pneumatic devices; or a Design Centre where students do their planning and problem-solving. You might even want to consider an Unbuilding Centre for students to disassemble, sort, and label the parts from wind-up toys, clocks, or unwanted small appliances.

STORAGE AREAS

You may wish to provide specific storage areas for tools and materials with the work surfaces conveniently located elsewhere. Placing tools in a box is not a good idea. Tools placed on tool boards mounted on perimeter walls are not only convenient for students, but allows the teacher to make a visual check for missing equipment at the end of each classroom period.

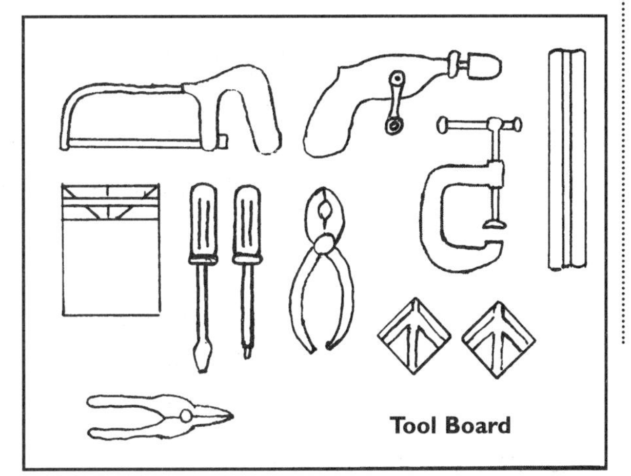
Tool Board

A helpful approach is to use labeled transparent containers. Items can be seen and identified, assisting students with the correct naming of parts, while contributing to an organized storage area.

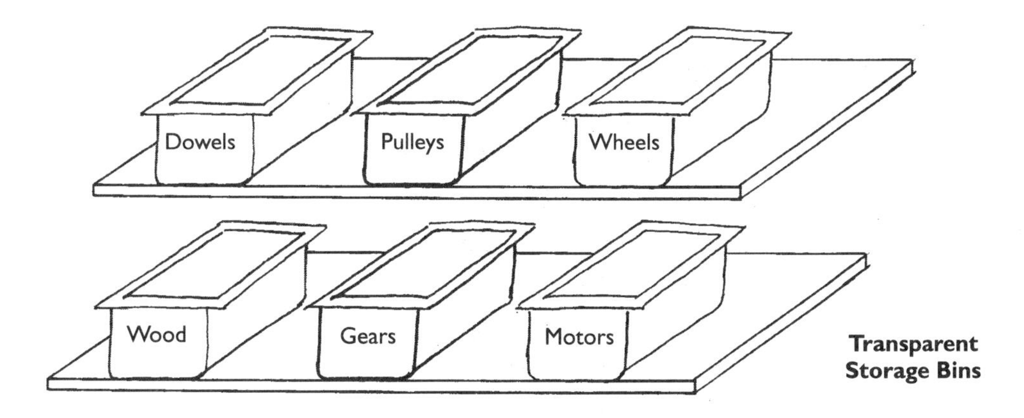

Transparent Storage Bins

WORK SURFACES

You can never have too many work surfaces in a classroom. With group work, students need space to be creative. Commercially-available portable benches, available at local lumber and hardware stores, can be purchased when space is minimal. Although these do not have a very large work surface, they can be collapsed and stored away when not needed. Hollow-core slab doors purchased from lumber suppliers make good work surfaces. These doors can be clamped on classroom desks and stored when not needed. If these doors are used for permanently-located work surfaces, they can be fastened on filing cabinets, clamped between portable benches, or have permanent legs attached to the under side of the door. Local secondary school woodshops could be asked to make workbenches for you as a project. Standard classroom tables also make good work surfaces.

METRIC MEASUREMENT

The measurements provided in this resource are metric, as far as possible. You will occasionally encounter Imperial measurements, particularly where they concern the sizes of materials and tools available at lumber yards and hardware stores. This is for your convenience when purchasing these items, because the construction industry, to a large degree, has not yet fully adopted the metric system. Some of the conversions (approximate) used in this resource are summarized here.

Sample Equivalent Table

Imperial Measures	Metric (approximate)
3/16 inch	5 mm
1/2 inch	12.5 mm
3/4 inch	20 mm
1 inch	25 mm
1 1/2 inches	38 mm
2 inches	50 mm

Note: Throughout this resource measurements are given in mm (and in rare instances in m). To convert mm to cm, divide by 10.

Section 3

BASIC TOOLS OF TECHNOLOGY

The following tools represent a basic tool set for most technological activities in the early grades. These tools are beyond what would normally be found in most classrooms. For instance, pencils, rulers, scissors, and staplers are all tools found in most classrooms and would likely be used as well in the designing and building of solutions to many challenges.

There are several other tools that would be used in specialized circumstances, but most often in the later grades when the complexity of the solutions increase and the ability levels of students mature. Hammers, screwdrivers, pliers, and wire cutters are tools that are necessary for many of these special applications. These and others are described in the next section as specialized tool sets.

Junior Hacksaw: This saw, used to cut 10 mm square strips of wood is the safest and least expensive of the available saws. It is designed to fit comfortably into student's small hands. It comes with both wood and metal cutting replaceable blades. The smaller toothed metal-cutting blades are easier for students to manipulate. This saw should be used with a Bench Hook.

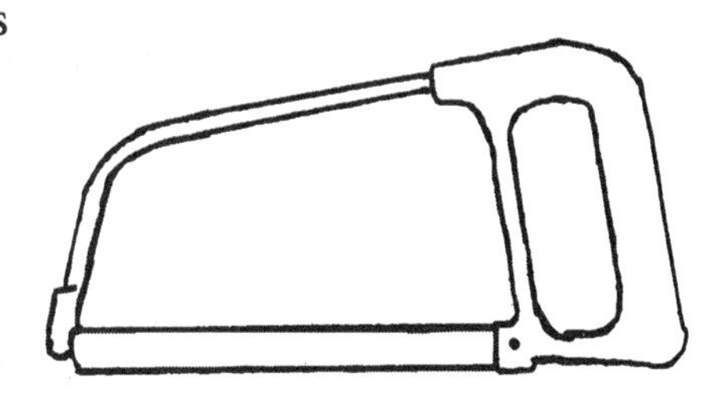

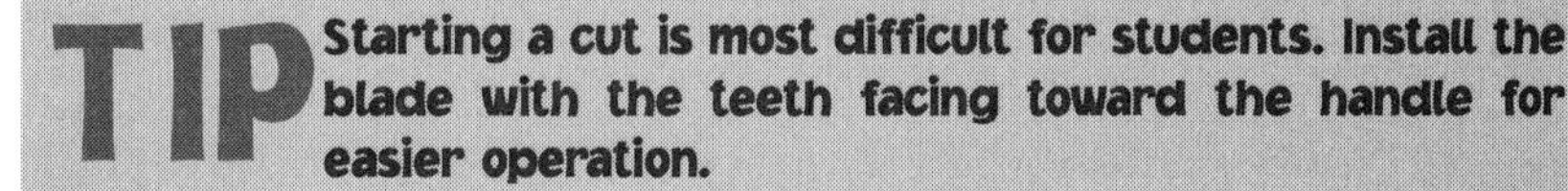

TIP **Starting a cut is most difficult for students. Install the blade with the teeth facing toward the handle for easier operation.**

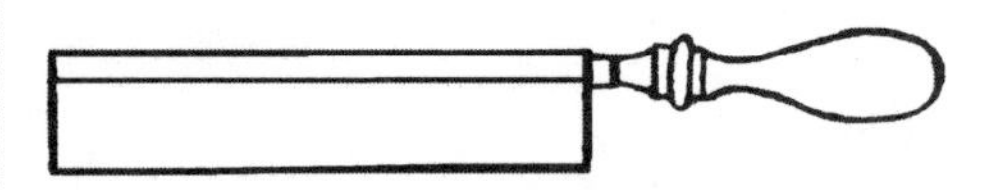

The Dovetail Saw: The Dovetail Saw is more expensive than the Junior Hacksaw, but sturdier. It will have to be sent out for sharpening on occasion. It is also used with the Bench Hook.

Bench Hook: The Bench Hook is for use with the Junior Hacksaw to hold the wood securely while cutting. The bottom cleat allows the Hook to rest firmly against the edge of the work surface. The top cleat often has a groove cut in it to hold the square wood. Saw cuts in this top cleat allow for accurate 90 and 45 degree cuts. These can be purchased or made in school woodshops. Use a C-clamp to secure the Bench Hook to the work surface. **Note:** See Appendix G for a drawing of a Bench Hook.

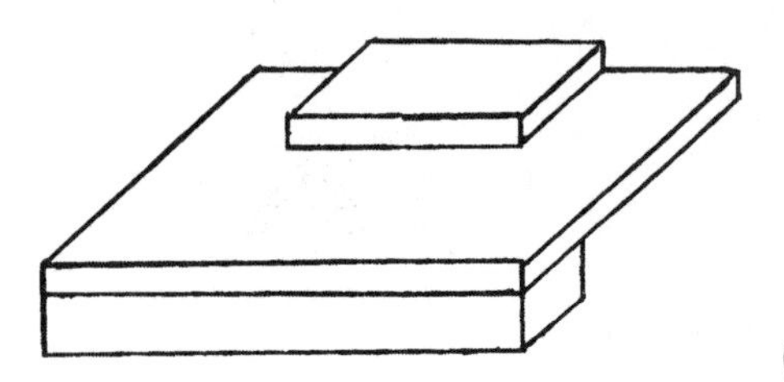

TIP **The bottom cleat can be made with two small blocks in each corner of the Hook surface to allow it to fit against circular work surfaces.**

SAFETY FIRST!
Avoid purchasing Hand Drills with open gear mechanisms in which fingers can be pinched.

Hand Drill: Hand drills are used to drill holes in wheels, pulleys, gears, and wood pieces. A chuck on the end of the drill is adjustable to accept drill bits up to 6 mm (or closest equivalent). There are several styles of hand Drills available, but the pistol-grip drill as shown is much easier to use than drills that resemble hand mixers. All drilling should be done with a scrap piece of wood beneath the piece being drilled to avoid holes being drilled in the work surfaces (desks etc.).

Drill Bits: The drill bit, sometimes called a "twist" drill, comes in a variety of diameters. Drill bits must be fastened securely in the 3-jaw chuck of the Hand Drill.
Bits must rotate clockwise to cut the materials being drilled.

Corner Joiner: The Joiner is used with wood strips and triangular cardstock "gussets" to make a secure corner joint. A cardstock gusset, with glue on the surface facing up, is first placed in the joiner. Wood strips are then placed on top of the glued gusset in the slots, provided in the joiner, to make 90, 60, 45, and 30 degree angled joints. A second gusset is glued on top of the joined pieces to make it more secure.

Gusset: A gusset is a small cardstock triangle of various sizes used to reinforce wood strips joined to one another. **Note:** See the section on Manufactured Materials for a description of gusset applications and Appendix E for a Gusset template sheet that can be photocopied on Card Stock.

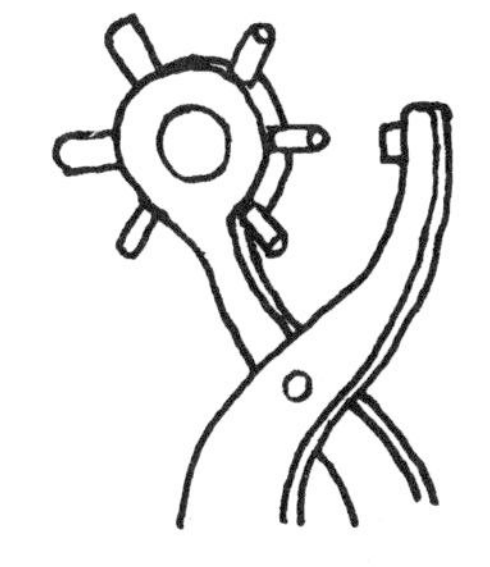

Hole Punch: The hole punch is used to put holes in gussets for dowel rod axles and shafts to rotate freely within. Single-hole punches can be used, but are limited in the variety of hole diameters required for dowel rods, art straws, or wire that may be used as axles and shafts. Heavy-duty punches are available to punch holes in Popsicle sticks or tongue depressors. A variation of the hole punch is the Paper Drill that has variety of removable cutter tips.

TIP **For accuracy, punch several gussets together at the same time. Glue several layers of gussets together for strength.**

SAFETY FIRST!
Only the capable should be allowed to use utility knives.

Metal Safety Ruler: The Metal Safety Ruler has all the measurement features of classroom rulers, but is designed with a groove in its top surface for fingers to rest safely out of the way when used with utility knives or rotary cutters. Its bow-like shape also helps to keep the ruler from moving around when drawing lines or cutting materials.

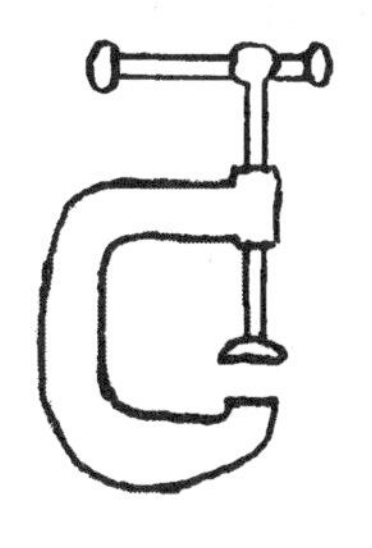

The C-Clamp: The C-Clamp is used primarily to hold the Bench Hook to the work surface while cutting wood strips. It can also be used to hold pieces of wood, or other materials, to the work surface while drilling and shaping or for clamping glued pieces together. The 75–100 mm (3–4 inch) clamp is the most common size used, but you may need a larger clamp if the work surface is thicker.

TIP **Clamps have a tendency to mark the wood when tightened. To prevent this, place a scrap piece of wood between the clamp and materials being clamped.**

Section 4

SPECIALIZED TOOLS OF TECHNOLOGY

The following tool sets are categorized as "specialized", which simply means they are designed for technological applications of a more specific nature. As the range of complexities in chosen solutions to a challenge increases with the maturing abilities of your students, these tools make specific tasks easier. For instance, when students want to add control to their devices, several tools designed to work with electrical materials are required.

SAFETY FIRST!

Low-temperature glue guns produce a weaker bond, but are safer to use, because they operate a temperature that will hurt, but not burn. Safety goggles should always be used with glue guns. You can identify low temperature glue sticks because they are oval rather than round in cross-section. Be sure glue guns are purchased which accept the oval-shaped glue sticks.

Table Vice: A Table Vice is used to hold wood strips, dowel rods, or other materials, while cutting, drilling, or shaping. The vice clamps to a table or other work surface and is available with a fixed or swivel base. The vice is also handy as a clamp for gluing pieces together and assisting in making sharp bends in wire or metal pieces.

TIP **L-shaped metal inserts or pieces of wood should be made to cover the jaws of the vice to protect the material being worked on from being marked.**

Glue Gun: The Glue Gun is not essential for most technology applications and should only be used to speed up project assembly and when gluing non-porous plastic materials. Be sure to use low-temperature glue guns. See the Safety First! tip.

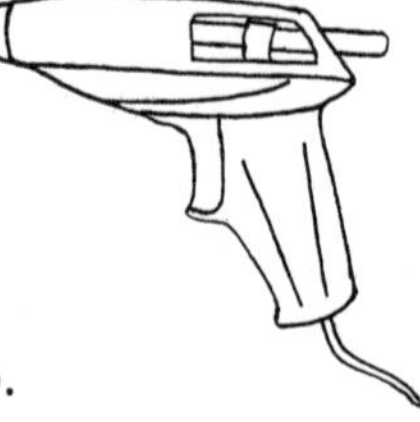

Hammer: The Hammer comes in a variety of sizes, but a smaller lighter size is recommended for small hands. Although there are many head shapes, the claw hammer, with its ability to be used as a lever to withdraw nails, is best for general use. With many applications using wood strips, nails would likely split these strips. The hammer then, is seldom used for many of these applications. Students do however, like opportunities to drive nails. Bring a log section into the classroom and let them practice.

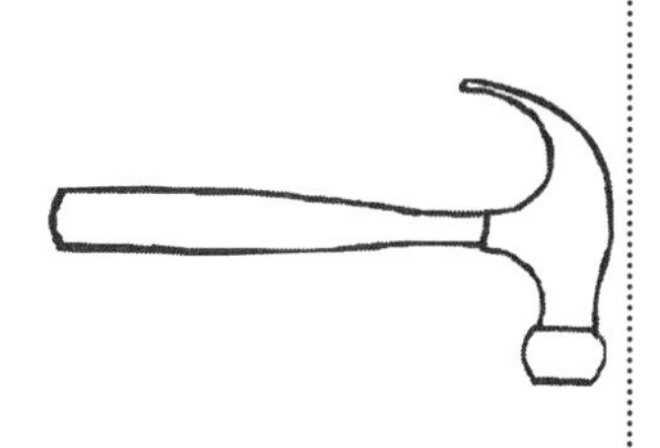

TIP Discourage students from "choking" the hammer, i.e. holding the handle too close to the head. They tend to do this because it gives a feeling of control, but results in a loss of force. The hammer should be held at the end of the handle.

SAFETY FIRST!

Goggles are recommended when using hammers.

Nails: Most nails are either "common" or "finishing" and come in a variety of lengths. Common nails have larger heads.

Screwdriver: Screwdrivers come with a variety of head drivers and shaft lengths. The most common head drivers are: A) Slot, B) Phillips, and C) Robertson. The latter two are easier to use because the screwdriver does not slip off the head of the screw. The 150 (6 inch) screwdriver is appropriate for most applications.

A B C

Wood Screws: Wood Screws also have a variety of head shapes with the round or oval head used to remain above the wood surface for decorative purposes. The flat-head wood screw is for applications when the screw is to be set below the wood surface. Wood Screws are also available in steel, zinc-coated steel, and brass.

TIP Screwdrivers must be held in line with the wood screw. Screws are inserted with a clockwise turn and removed with a counter-clockwise turn.

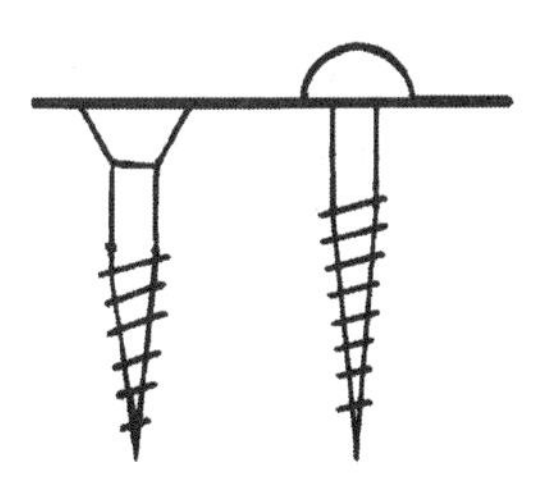

TIP **Pliers are also a type of lever, the closer to the pivot, the stronger the hold (or cutting power).**

Needle-nose Pliers: Needle-nose pliers are particularly useful when working with electrical wire (forming wire loops) and devices, and for reaching into areas that would not be possible with regular pliers. Some of these pliers have a cutter feature to use for cutting light electrical wire.

Side-cutting Pliers: Side-cutting pliers are designed for cutting wire materials. They have a "Mechanical Efficiency" that will easily cut heads off nails or wire up to coat-hanger diameters.

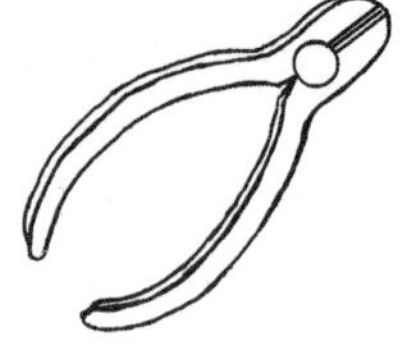

Wire Strippers: Electrical wire usually has an outer insulation coating over an inner copper core. To make contact with electrical devices, a short piece of the copper core must be stripped of its coating. Some types of Strippers, similar to the one shown, remove the right amount of coating for most applications. With other less expensive types, about 12–15 mm have to be removed by hand.

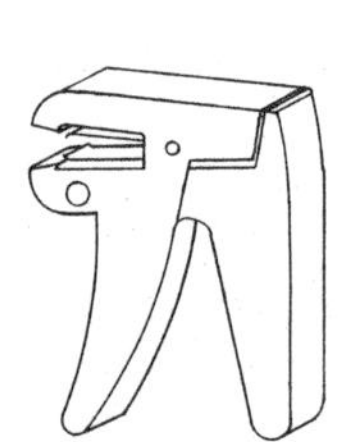

Snips: Snips are a stronger version of scissors used to cut thicker material such as tin, plastic sheets, and plastic tubing. The handle length coupled with the lever action of the jaws of the Snips, produces a powerful cut.

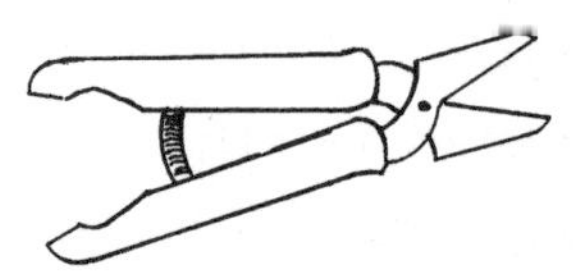

Utility Knife: The Utility Knife is used for making long straight cuts on cardstock and paper. It is also handy for cutting openings in cardboard boxes. The Utility Knife must always be used with the Metal Safety Ruler that keeps fingers out of the way of the razor sharp blades.

Files: Files are used to remove small amounts of wood to make pieces fit together better. Several shapes are available (flat, half-round, round). The round file is useful for slightly enlarging holes in wheels when a loose fit is required on their axles. Files have a "tang" on their ends to hold wood or plastic handles. Handles should always be used.

SAFETY FIRST!

SNIPS have a powerful jaw action. Caution students to keep their fingers well back from the cutting edges when using this tool.

Before permitting students to use UTILITY KNIVES, be sure they are capable. You will need to individually assess each student's ability to safely use these and any other potentially hazardous tools. If in any doubt, have the student work with a capable partner or under personal supervision until you feel that person has gained the necessary skill and experience.

FILES are very brittle and will easily break if used for anything other than filing.

Section 5

MATERIALS

At all grade levels, students involved in creative problem-solving will need a range of materials to consider for their challenge. Several factors such as cost, storage space, and the grade level abilities of the student will dictate what materials are most suitable for the problems to be solved. Junk (recycled) materials are available at no cost. A "Please Save For Me" note (see Appendix A) could be sent home to parents/guardians requesting a limitless supply of these items. Several "throw-away" items can also be requested from local retail shops and businesses in your area. Manufactured materials have a cost involved, but many can be re-used to keep costs to a minimum.

As students become involved with materials, their experiences will determine which is most appropriate for solving their problems. For instance, they will find that card stock is more rigid than paper for structures. Similarly, wooden or plastic wheels are easier to secure to axles than tin or cardboard alternatives. The following is a partial list of the most common junk and manufactured materials to consider.

SAFETY FIRST!
Extreme care must be taken when having students handling any containers that may have been used for food. These must be thoroughly cleaned to avoid the possibility of salmonella food poisoning.

JUNK MATERIALS

These materials are ideal for challenges in solving technological problems in the early grades. Each will provide a beginning experience in working with materials even though for the most part they can only be used to make representative models of structures and devices. Some of these materials, used in conjunction with manufactured materials, allow for the design and assembly of working models. Useful junk materials include, among many others: film

SAFETY FIRST!

Electrical cords on appliances must be cut off before students begin any disassembling.

Do not have students disassemble TV sets or computer monitors. The stored electrical charges in some component parts can give a dangerous shock. Also, mishandling TV tubes can cause them to implode.

canisters, drink cans, metal and plastic lids, cardboard boxes, cardboard tubes, spools, plastic bottles, and styrofoam and aluminum food trays and boxes. To assist with the development of organizational skills, each of these materials should be classified and stored in appropriately labeled containers.

In addition to the junk materials listed above, you might consider asking for old, broken, or unwanted small appliances. Items such as mixers, blenders, toasters, radios, and small power tools can be disassembled at an "Unbuilding Centre" and each of the parts sorted and stored in small plastic margarine containers. Apart from the experiences gained in taking things apart, students can learn how to develop their own classification system of similar parts. Several of these parts will also find a use in their design solutions.

MANUFACTURED MATERIALS

Over the years, manufacturers have produced materials designed for specific technological applications in elementary schools. These materials, available from educational suppliers, allow students to design and make structures, mechanisms, and devices that are working models of real-world applications.

1. WOOD STRIPS

The selection of wood used in the classroom should be considered carefully. Without the availability of machines and carpentry tools that are needed to cut and shape materials in most classrooms, the use of large dimensional materials should be discouraged. Similarly, hardwoods are more difficult to cut than softwoods and should also be avoided.

Small dimensional softwood (basswood or pine) strips (5, 8, and 10 mm square stock) usually in lengths of 400 mm, are available from educational suppliers. These small wood strips are easily cut and assembled to make two and three–dimensional structures and devices.

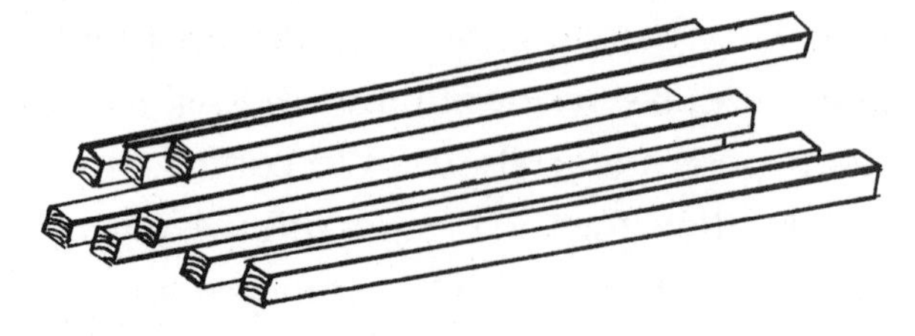

2. DOWEL RODS

There are several diameters and lengths of dowel rods available, with 5 mm diameter being the most often used. This diameter conforms to most of the pre-drilled holes in supplier's wheels. Pulleys and gears often come with 4 mm pre-drilled holes. Dowel rods, to suit this diameter, are also available or the holes in the pulleys or gears can be drilled to suit 5 mm dowels. Dowel rods are used as axles for wheels, shafts for pulleys and gears, and pivots for hinged joints.

3. GUSSETS

Gussets are card stock triangles used to secure the ends of wood strips rather than having to use nails. When glued to both sides of the surfaces to be joined, they make a relatively strong joint. Gussets can also have a hole punched in one of the triangle corners to hold dowel rods for axles or other rotating devices. These can be purchased or if your photocopy machine accepts card stock, can be duplicated using the black-line master included in Appendix E.

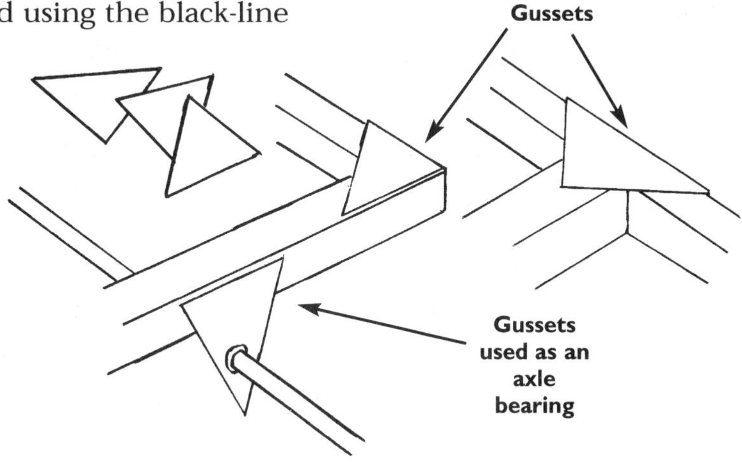

4. ART STRAWS

Art Straws are longer than standard drinking straws and made of paper making it easier to fasten them together with glue and other fastening devices. Available in both regular and jumbo sizes, the jumbo size Art Straw has the advantage of allowing 5 mm dowel rods to rotate within when used as axle bearings. The regular Art Straw will fit inside the jumbo when extra strength is required. These straws are easily cut and shaped to make simple structures.

5. WHEELS

Wheels are available in cardboard, plastic, and wood. Wheel diameters are 25 mm, 38 mm, and 50 mm (for conversion to inches, see page 22). Normally these come with a 5 mm drilled hole. Others have holes that are slightly smaller and have to be drilled to suit the dowel sizes. The plastic and wooden wheels are best for functional (working) models of vehicles and other devices. Cardboard wheels are best for simulated (non-working) devices although several of these wheels can be glued face-to-face to make a thickness suitable for functioning models. Other materials such as film canisters, jar lids, thread spools, and drink cans can be used as wheels or rollers. On-centre holes will have to be drilled in these to fit the 5 mm dowel diameters used for axles.

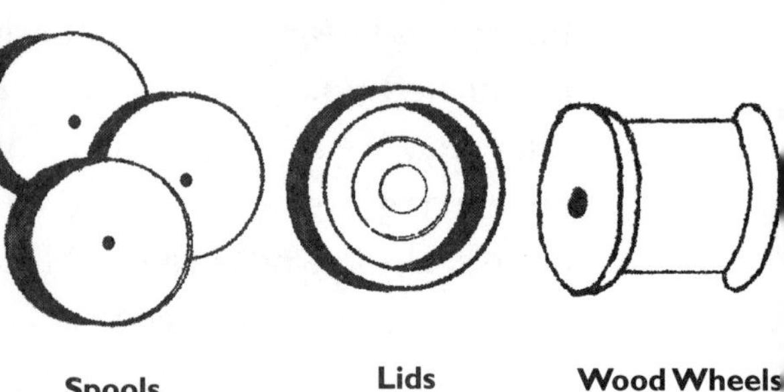

6. PULLEYS

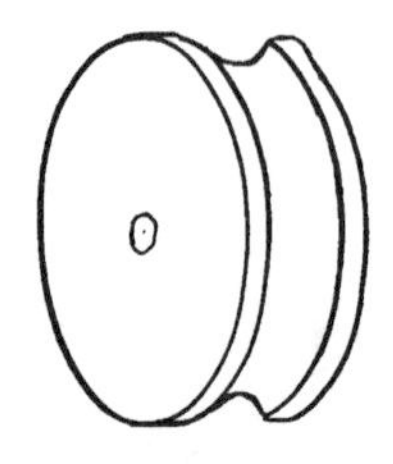

Plastic Pulley

Plastic and wood Pulleys with pre-drilled holes to fit on dowel rods are available. The pulley diameters are 25 mm, 38 mm, and 50 mm (for conversion to inches, see page 22). Normally these come with a 5 mm drilled hole. Others have holes that are slightly smaller (4 mm) and have to be drilled to suit the 5 mm dowel. Small pulleys with 2 mm holes, that fit on the end of the shafts of small electric motors, are also available. Pulleys can be made by sandwiching a small wheel between two larger wheels. Thread spools also make excellent drum-type pulleys.

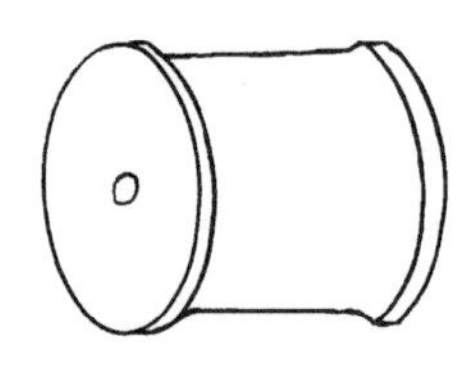

Thread Spool

7. GEARS

Plastic gears, with pre-drilled holes to fit tight on 4 mm dowel rods, come in a variety of types and sizes. Their pre-drilled holes can be easily enlarged with a 5mm drill bit to fit on the standard 5 mm dowel. Standard spur gears come in 25 mm, 38 mm, and 50 mm sizes. In addition to these, Bevel gears, designed to run at 90 degrees to one another; Rack gears for linear movement; and Worm gears that are spiral in design, are available (see pages 51 and 52 for a diagram of these gear types). Some challenges might suggest that students make their own gears. Plastic gear jigs are available that allow students to make 8, 12, 16, and 24-tooth gears from cardstock disks that sandwich 5 mm dowel rods or wood strips. See the section on Machines and Mechanisms for a description of these gears and their applications.

8. ELECTRICAL COMPONENTS AND SUPPLIES

With so many possibilities available for students with appropriate skills, understanding, and experiences, adding energy and control to their design solutions is made possible with a full range of electrical components and supplies. These include Battery Holders, Buzzers, Motors and Motor Holders, Switches, Bulbs and Bulb Holders, and wire to connect these components in a variety of circuits. Miniature Solar Panels also provide an alternate energy source.

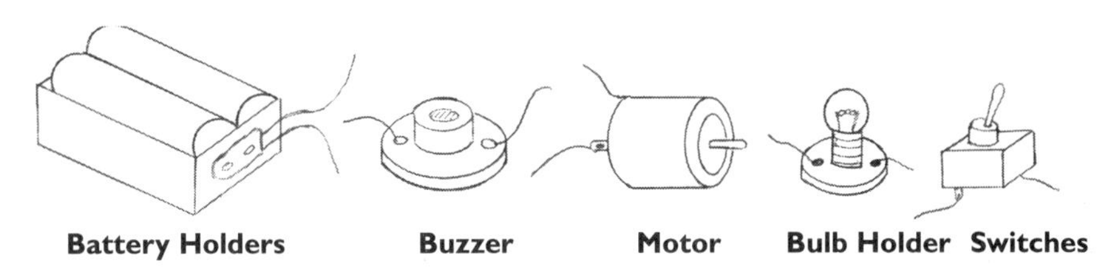

9. SYRINGES AND TUBING

Another form of control that students can add to their solutions to challenges is with pneumatics or hydraulics. Syringes and tubing provide the means to activate their devices. Syringes come in sizes of 5, 10, 20, 30, and 60 ml, with the smallest three sizes providing most of the control needed. Tubing comes in rolls and is a flexible PVC material that fits all sizes of syringes. When students need to control the movement of more complex devices, Connectors, Tees, and Valves are also available. A pneumatic system is easily converted to hydraulics by introducing water into the system. Although sometimes messy when using water, it does eliminate the compression of air in the pneumatic system to give more positive movement.

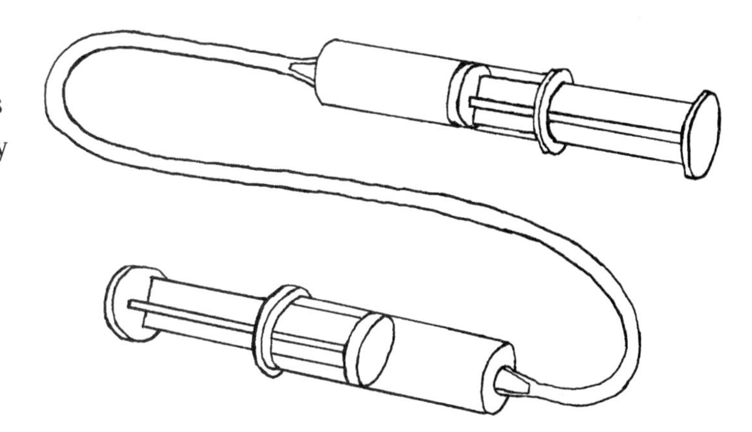

Section 6

TOOLS AND MATERIALS APPLICATIONS

One of the difficulties facing the untrained student (and quite a few teachers) is the range of possible uses of the materials and the tools used to assemble these materials into a variety of product solutions for each challenge. It's not unusual for students to need to experiment with these materials to find the advantages and limitations of each. Paper, cardstock, scissors, and glue will provide lots of possibilities for simple structures, but when used with wood strips or even Popsicle-sticks, design flexibility and stability can be added. Many structures such as houses or castles may be non-functioning models, but with the addition of a hinged door on the house or a drawbridge on a castle, the mechanisms required for these have added a new range of experiences for students. Adding wheels, axles, and bearings to vehicles will introduce motion, but a battery, switch, and motor that drives a pulley mounted on the axle, controlled motion becomes an extension possibility as students gain experience in experimenting with these materials. The use of different kinds of tools for these applications will also broaden the students' skill development.

BASIC TOOL TECHNIQUES

Before untrained students can safely and accurately use hand tools to measure, cut, and assemble materials, they should be given an opportunity to use these through a couple of introductory projects. The following frame can be used for pictures, artwork, or a seasonal greeting. It is a model that will introduce students to the need for accurate measurement, to the techniques of sawing, and to the assembly of materials.

CHALLENGE

Make a square frame measuring 150 mm × 150 mm outside to outside measurements. (use 10 mm × 10 mm square wood strips)

Note: The frame is made with two front and back pieces that are 150 mm long and two side pieces that are shorter by two thicknesses of the 10 mm square strip. The most common error made by some students is forgetting to shorten the two side pieces when outside measurements are important.

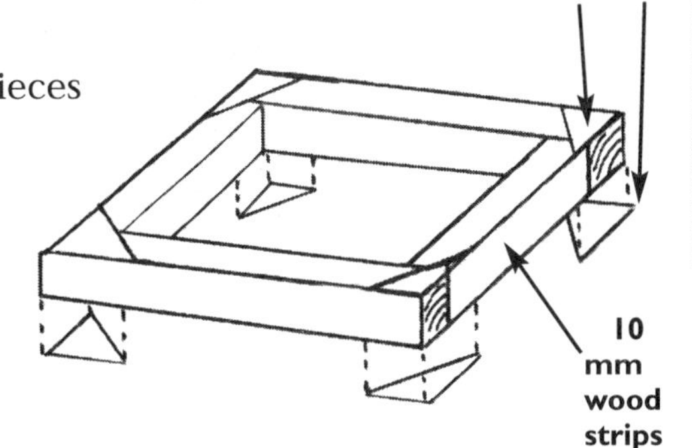

Measuring the Pieces

Students should be encouraged to make accurate measurements. A sharpened pencil will assist with this accuracy and when measurements are accurate, the parts go together well. Set the 0 (first graduation) in line with the end of the wood strip as shown below and holding the pencil away from you, mark a short line.

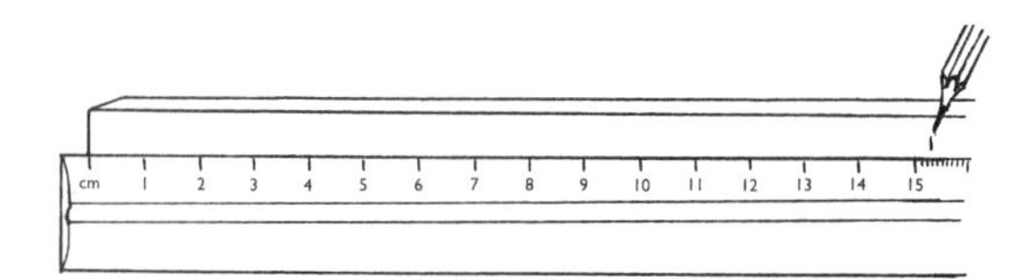

Sawing the Pieces to Length

Students should be encouraged to saw on the "outside" of the line. Cutting "on" the line will make the piece short because of the width of the saw cut. Clamp the Bench Hook to the table or bench surface to keep it from moving around while sawing. The wood strip to be cut, should be placed against the top cleat on the bench hook and with the marked line at the end of the cleat. The end of the cleat is used to keep the saw blade square to the marked line.

To start the cut, tilt the saw slightly, as shown below, to begin the cut on the front edge of the wood strip. Since all cutting is done on the forward stroke, downward pressure should be on this stroke only.

Drilling the Pieces

Drilling holes requires some care to prevent holes ending up where they shouldn't be (in desk or tabletops) and to ensure parts drilled are properly aligned. Pieces to be drilled need to be clamped on top of a scrap of wood to protect work surfaces or clamped in a vice.

Assembling the Pieces

Assuming the pieces have been cut accurately, the task now is to assemble the pieces so that the frame has square (90 degree) corners. Gluing the corners using a Corner Joiner, as shown below, will assist with it being square. Apply a thin coat of Carpenter's glue on the surface of a gusset and place it in the bottom of the Corner Joiner and with the glued surface facing up. Apply a thin coat of glue on the end of the short side piece and place the strips on top of the glued gusset. Complete the corner with a second gusset glued on top. Allow the glue to set a couple of minutes, remove from the joiner and repeat the process at each corner.

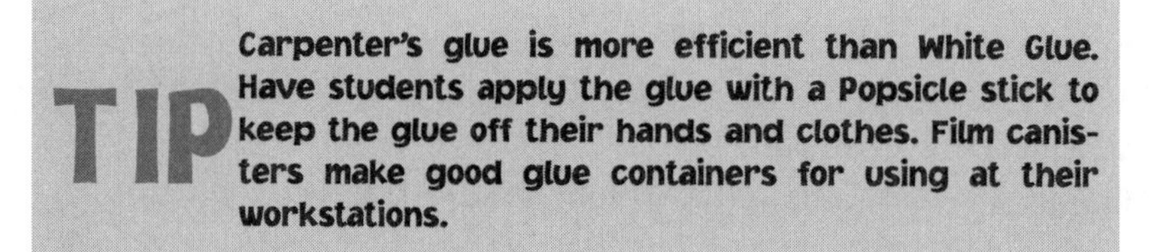

The following is an introductory project that can be used to develop the skills in drilling holes and give more practice in sawing and assembling materials. Students can also begin to understand line drawings.

CHALLENGE

Make a Pin Hinge with each of the hinge pieces cut and assembled using the dimensions shown in the following sketch.

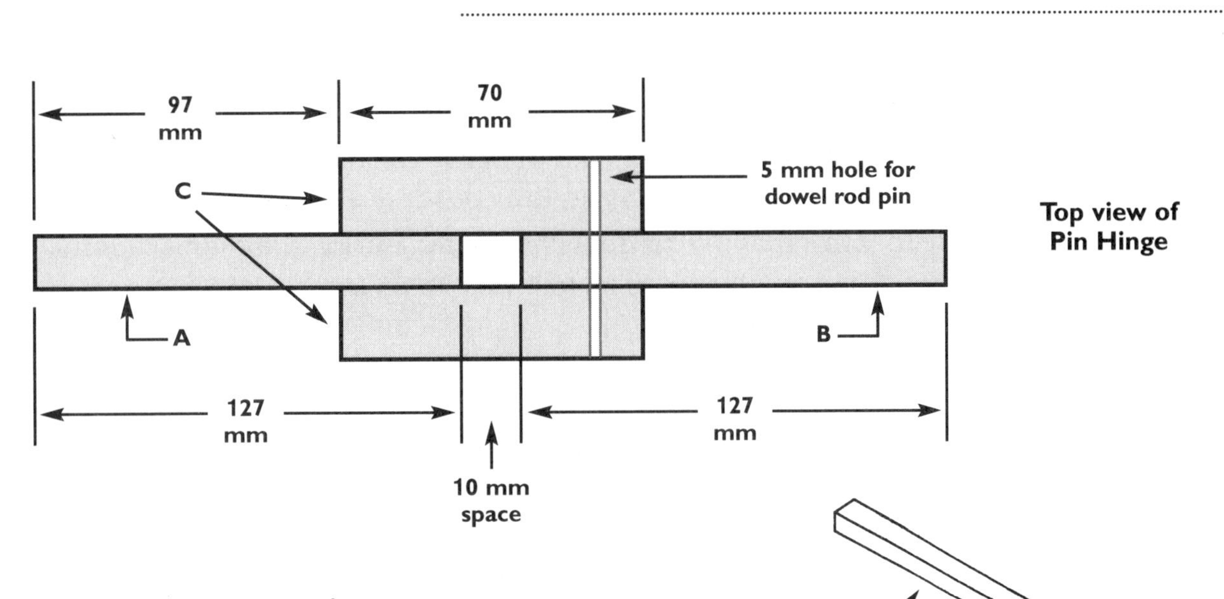

Sawing and Assembling

1. Cut the two 127 mm pieces (A and B) and two 70 mm pieces (C).
2. From the end of piece (A), mark a line on two opposite edges to locate the positions of (C).
3. Glue the two pieces (C) onto piece (A) at the marked line.
4. Use a C-Clamp to hold the pieces together until the glue dries.
5. Insert piece (B) between the (C) pieces and place the total assembly in a Table Vice and drill a hole through all three pieces.
6. Cut a 30 mm length of 5mm dowel rod and insert it through all three pieces before removing the C-Clamp.

Drilling

When drilling holes, the drill bit must be held 90 degrees to the surface and using light pressure. With hand drilling, holding the drill bit at 90 degrees to the surface can only be determined by eye. With practice, this gets better. When drilling through pieces without the use of a vice, always place a scrap piece of wood under the piece being drilled to protect the desk or work surface.

A
C
B

Drill through all the pieces together

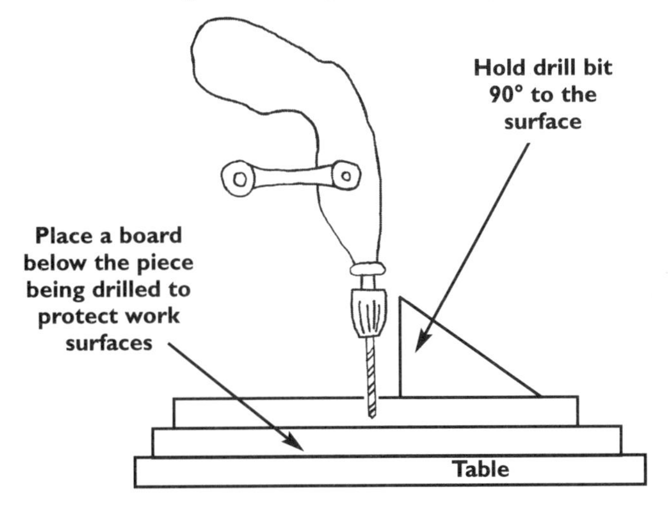

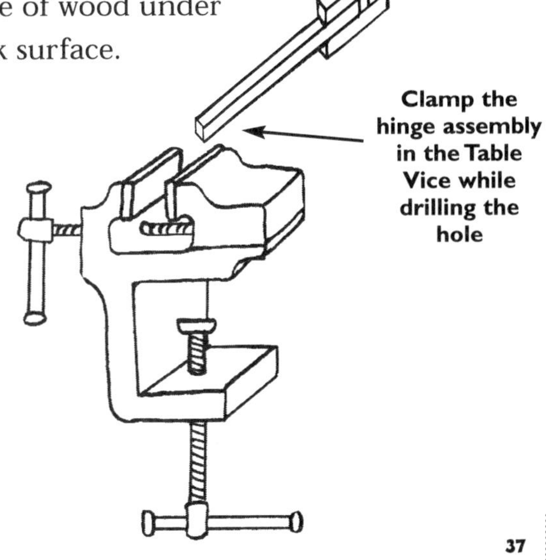

SPECIALIZED TOOL TECHNIQUES

The foregoing two challenges will provide students with an opportunity to practice some of the more common techniques using a basic set of tools that will be duplicated often during their application with future challenges. The following looks at some of the tools and techniques required when working with wire and electrical devices.

As mentioned previously, electrical wire comes with a non-conducting plastic coating. In order to make a positive connection to electrical devices, 12 to 15 mm of this plastic coating must be stripped off the end of the wire. Several types of Wire Strippers are available that make this easy (see diagram in the Specialized Tools section). The Wire Stripper shown below is an alternative type that is equally effective but requires some choice of holes to suit the wire diameter.

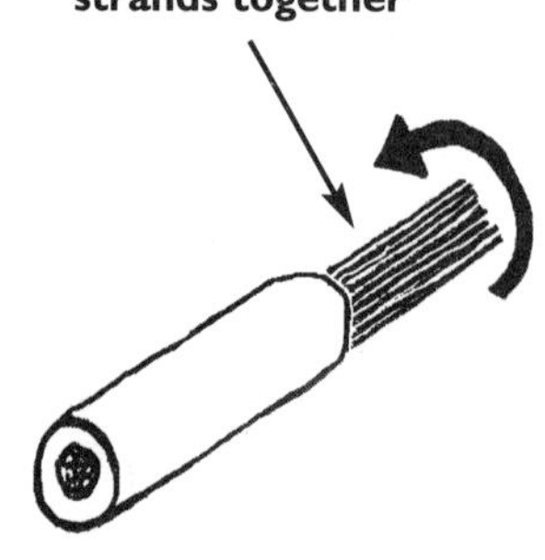

Some wire has a solid copper core, while other types have several small strands of wire. Stranded wire has to be twisted together after the plastic coating has been stripped to keep the strands from fraying and the loose ends touching the wrong terminal, causing a short circuit.

Once the ends of the wire have been stripped, Needle Nose Pliers are used to bend a loop in the wire ends to fit securely under a fixing screw provided on some electrical devices.

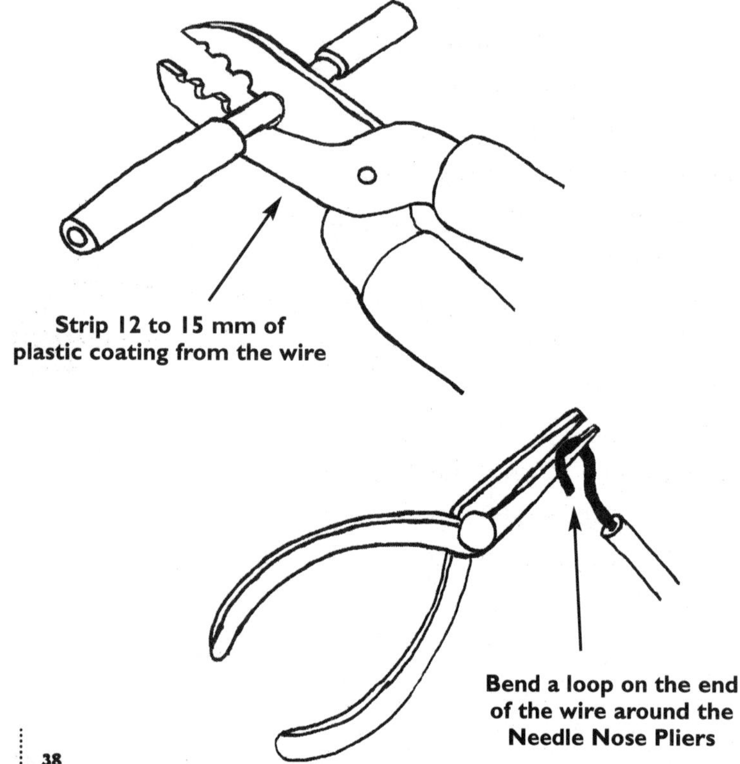

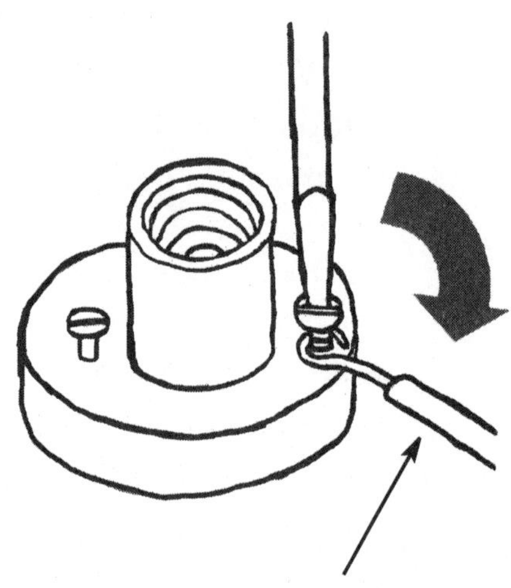

Section 7

Building Structures from Different Materials

There are basically two types of structures; Frame Structures and Shell Structures. Frame structures are made of separate parts called "members" that are joined together. A chair, a suspension bridge, and a house are examples of frame structures. Shell structures are not usually made up of separate parts but get their strength from the way they are shaped. A drink can, a dome, and a car body are examples of shell structures. Often, a structure consists of both a frame and a shell. For instance, the automobile has a "chassis" or "frame" that is an assembly of parts that in turn supports the car body (shell). The human body is another example with its skeleton frame covered by a skin shell.

Before students begin using materials to fabricate (forming and assembling) structures, they should have some knowledge of the load-bearing capacity of materials and their contribution to the strength of the structure. If a structure cannot withstand the internal and external forces that act on it, it will collapse. A chair, for instance, will collapse from the weight of the person sitting on it if was not made from appropriate materials and assembled in such a way as to provide resistance to the external force of the person's weight.

In the early grades, have students place a strip of card stock between two books spaced about 20 cm apart. Use coins or objects of similar weight as a load (external force). They will notice that the card stock deflects (sags) downward easily. Try gluing 3 or 4 layers of card stock together and place the same weight on this laminated strip. Is there a correlation between mass and strength?

A structure can be made stronger by changing the shape of the materials. Have students shape some card stock as shown below and place these shapes between the books at the same spacing. Which shape has the most strength? Have students look at the ends of a piece of cardboard. Which of the shapes below when sandwiched between two card stock sheets would resemble cardboard? Is it the mass or shape of the cardboard that gives it strength?

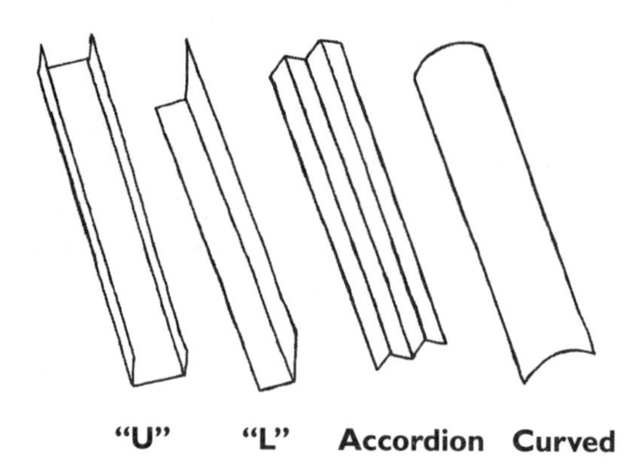

"U" **"L"** **Accordion** **Curved**

The strength of the frame structures students build will be a test of their understanding of capability of the the frame to support its own weight and the weight of the "load" that is put on it. Students will find that the triangle is a strong geometric figure and is often used to stabilize structures. To test the strength given by the triangle, have students make up a simple rectangle with strips of card stock secured at the corners with paper fasteners. The rectangle will easily distort into a parallelogram. Adding a diagonal strip as a brace will stop this distortion. Show students pictures of steel towers to see diagonal bracing in structures.

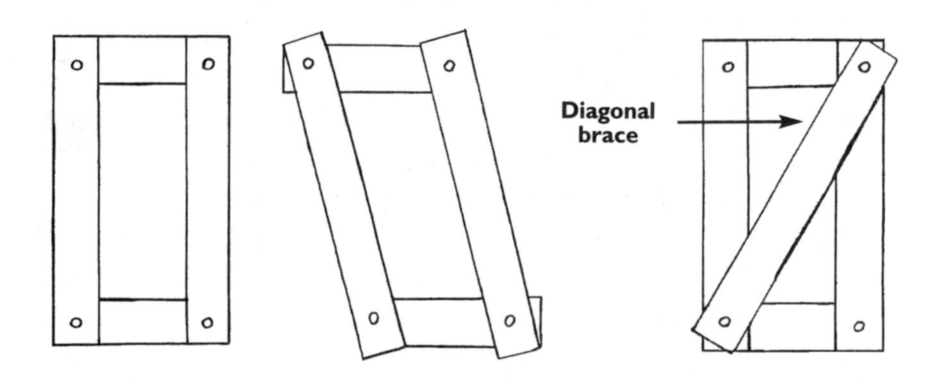

CARDBOARD MATERIALS

Just about anything the inventive mind of a student can imagine can be structured from paper, card stock, cardboard boxes, cardboard tubes, food trays, and a little bit of glue. A pair of scissors is the only tool required. These cardboard materials can be painted to hide the pictures, labels, and writing that are on their surfaces.

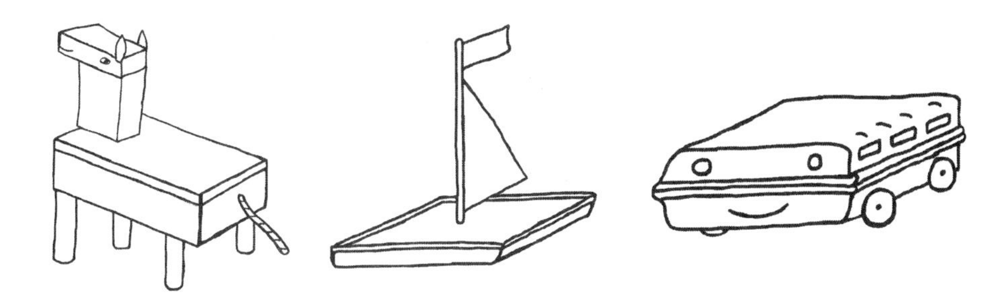

Making Boxes

Having students make boxes from card stock, or similar materials, promotes a simple measurement and layout skill and gives the design a new look. Let students begin by first taking a box apart. Have them first looking inside the box to find, and carefully unglue, the narrow tabs that may have been one of the few means of holding the box together. When apart, they will see that it is made from a single sheet of material not unlike the box layout shown below.

Have students make a "Story Box" and tell the class their stories about the images they have drawn on each of the outer surfaces of the cube. You can have students create their own layout or use the "Black-line master" (Appendix B) as a template for the cube. Once a desired layout has been made, scoring each of the layout lines with the point of a pair of scissors will allow each fold to be sharp.

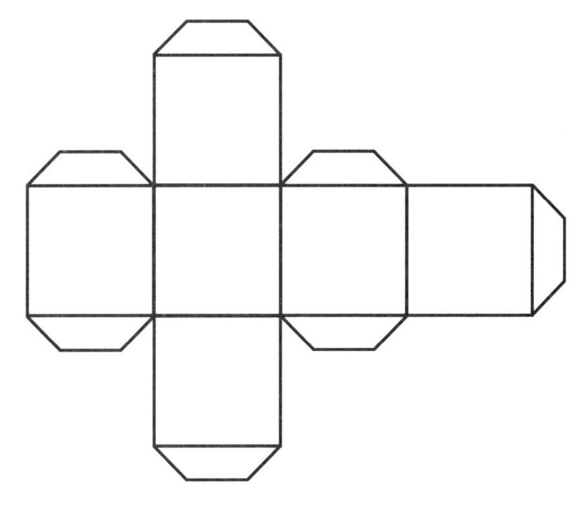

ART STRAWS

Art Straws can be used to create a variety of structures because of their simplicity to cut, bend, and fasten together. Both the regular and jumbo sizes can be used together for a variety of design ideas. Students will find that the regular straw fits within the jumbo straw making the straw stronger. Short pieces of both sizes can be used to join the ends of the other together by inserting and gluing the smaller size within the larger, or by gluing the larger on the outside of the smaller as a sleeve. Using cardstock gussets to join the end of one straw, to the end or middle of another, will add structural rigidity. Art Straws can also be woven together to make interesting patterns. Making towers or platform structures aids students to learn that the triangle is a strong geometric shape.

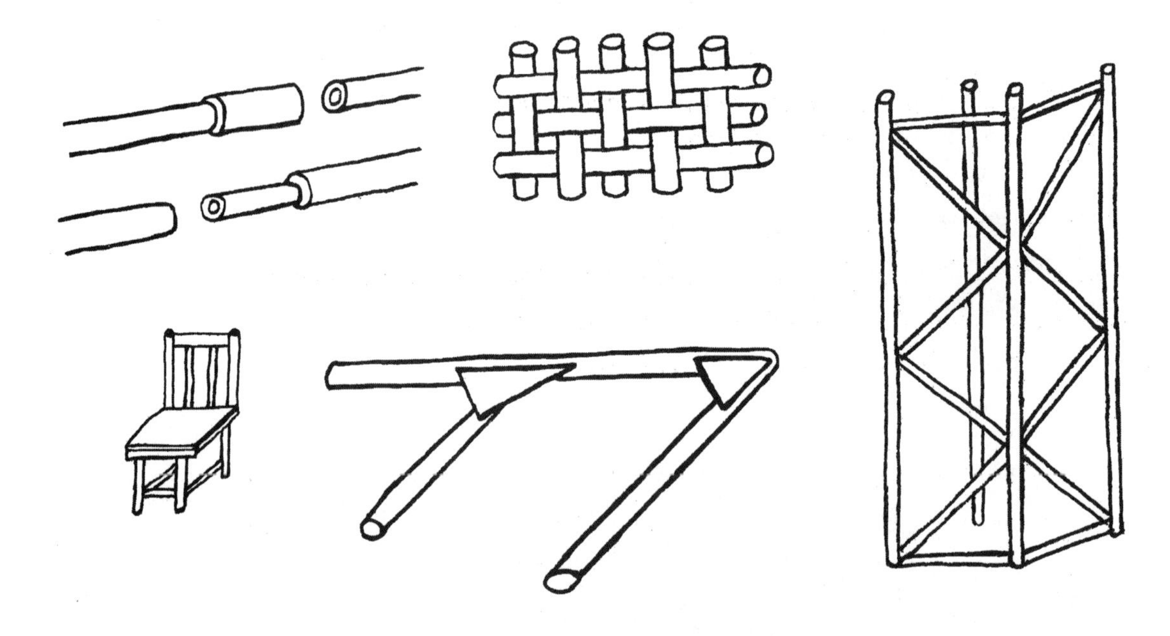

WOOD STRIPS

The use of wood strips will allow students to design and make just about any type of structure imaginable. The use of triangular cardstock gussets glued to the wood strips wherever separate members join one another, produces a relatively strong frame. Frames can then be covered with paper or card stock to give some realism to the structure.

A three-dimensional frame, which represents the basic frame for most structures, is made by putting two "flat" two-dimensional frames together with vertical members at the corners. Other three-dimensional frame structures representative of cranes, vehicle frames, or suspension bridges, also begin with this basic procedure.

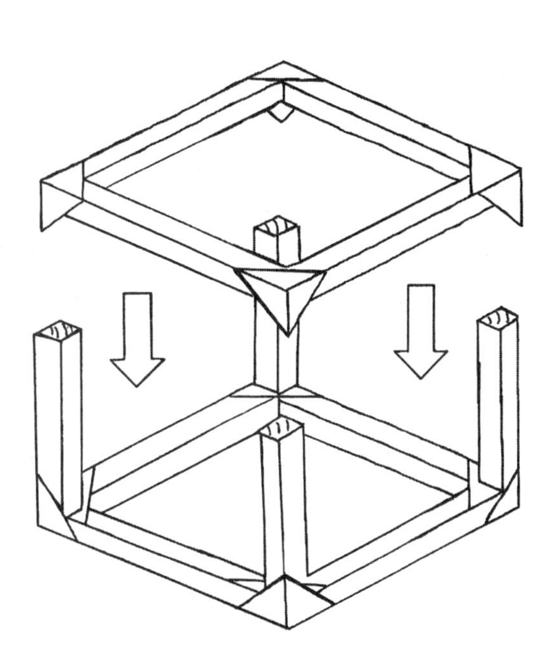

TIP Wood strips will split if students use nails to try and fasten them together.

As students become familiar with the procedures of assembling frame structures, their imaginations will allow them to produce some realistic yet challenging structures.

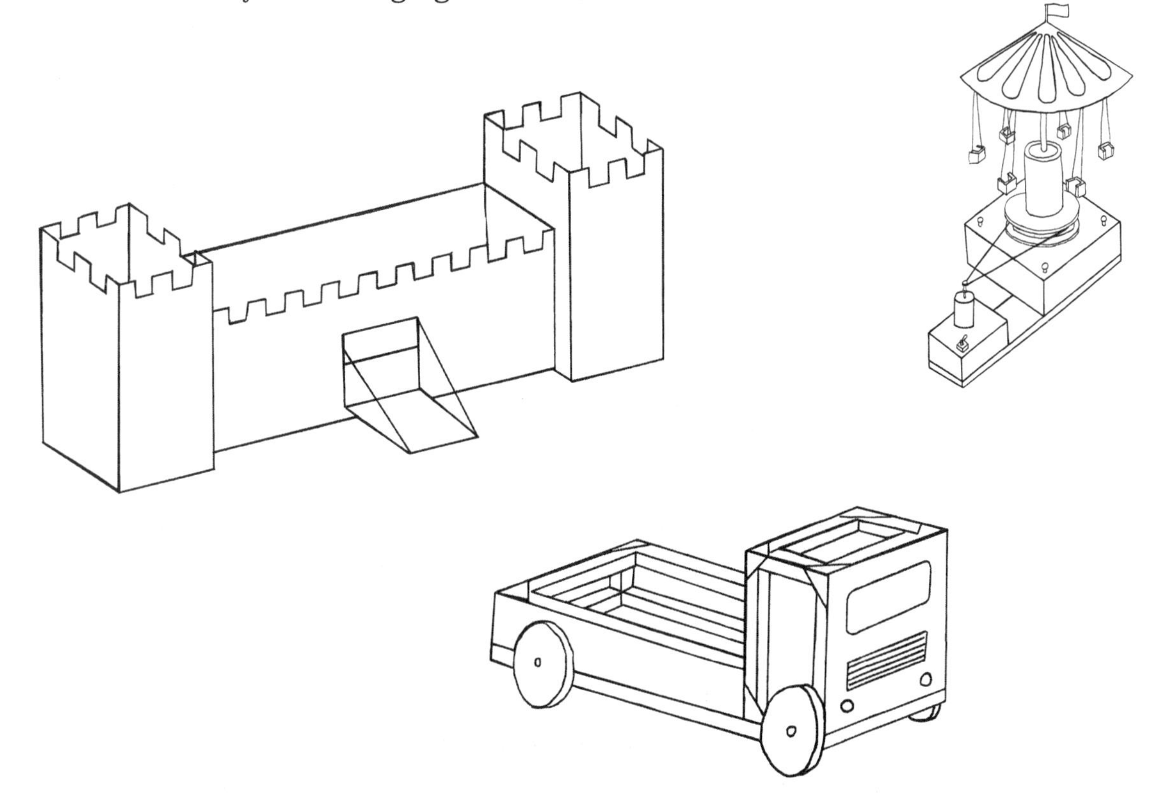

Section 8

Machines and Mechanisms

Any object that helps make our work easier is called a machine. We naturally think of the factory or shop as places where we would see machines, but some of the simplest machines are knives and axes, that can be used as tools. Within the factory, we see Drill Presses, Table Saws, Lathes, Robots, and other production equipment. In the transportation field, we see machines that range from the tricycle to the automotive vehicle. Around our homes we see computers, telephones, vacuum cleaners, and lawn mowers. All of these machines help us to work faster, easier, and they increase the amount of work we are able to do.

All machines are driven by some form of energy. The bicycle is powered by human energy; the automobile by gasoline; the pneumatic cylinder by air; and the windmill by the forces of the wind. Appliances around our homes need energy in the form of electricity to work. Earlier machines were driven by steam as the source of energy.

Each machine is also designed to incorporate "mechanical efficiency" (mechanical advantage). This is an attempt to provide less "effort" when a "load" is to be moved or lifted. A machine can have one or more simple machines as its working parts that are often referred to as "mechanisms." Several of these mechanisms might be working together within a machine to provide mechanical efficiency.

Please note: the following descriptions of levers, wheels, pulleys, gears, etc. are not intended to be an exhaustive coverage of the subject, rather are intended as a general overview of these mechanisms for purposes of their use in challenges.

LEVERS

The lever is the simplest and oldest of machines that is used to increase the mechanical efficiency in a mechanism. The teeter-totter or seesaw is an example of a lever. A carpenter's claw hammer used to remove nails or a screwdriver, used to pry lids off paint cans, are also levers. Around the home, we use lots of simple machines such as scissors, nutcrackers, and pliers; all of them levers and designed to make work easier. Have students identify first, second, and third class levers from several examples of levers.

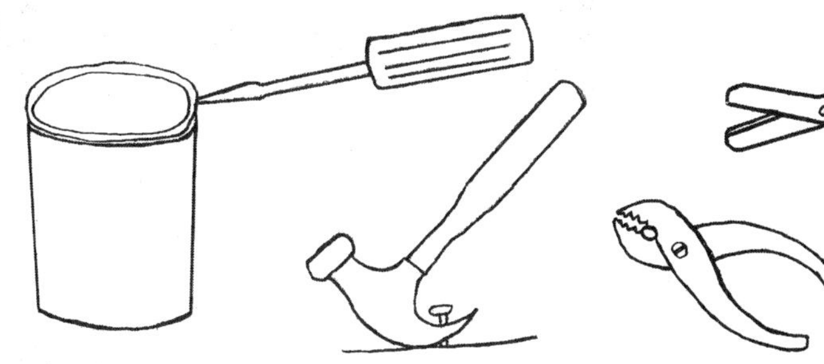

The improved mechanical efficiency of a lever is the result of the fulcrum or pivot being placed as close as possible to the load. The following outlines this basic principle.

1. The effort applied is equal to the load when the fulcrum is centred between the load and effort points.

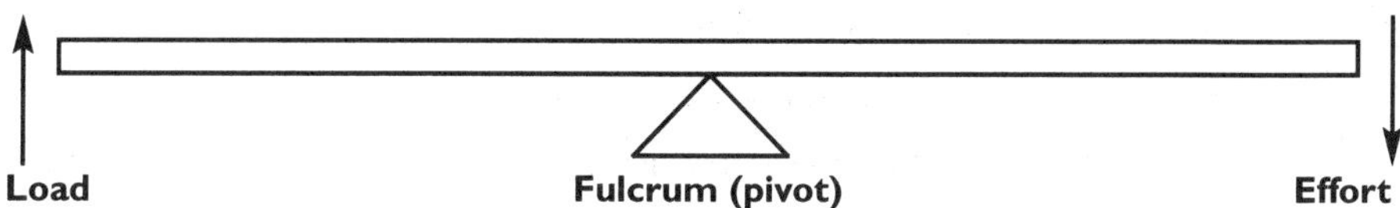

2. The effort applied is greater when the fulcrum is further from the load.

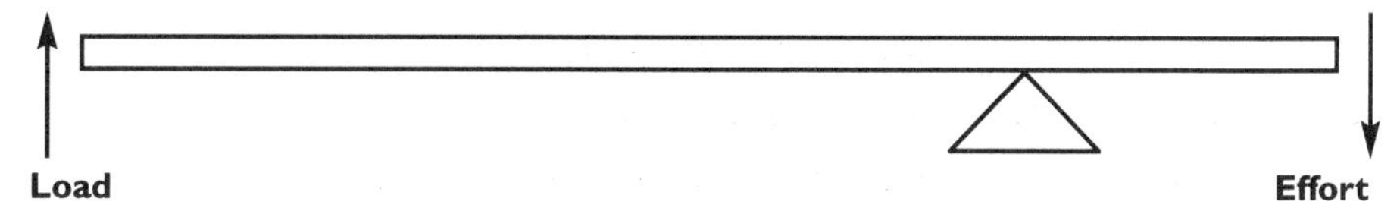

3. The effort is made less as the fulcrum gets closer to the load (greater mechanical efficiency).

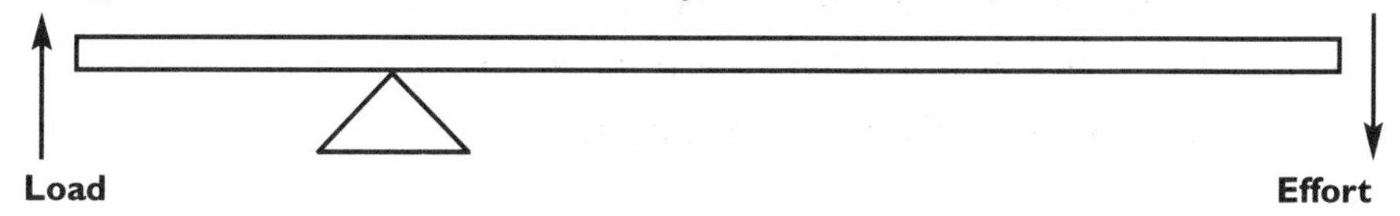

4. If load A and B are equal, there is less lift effort required to lift the load if it is positioned at B. This is the basic principle which gives the wheelbarrow its mechanical advantage.

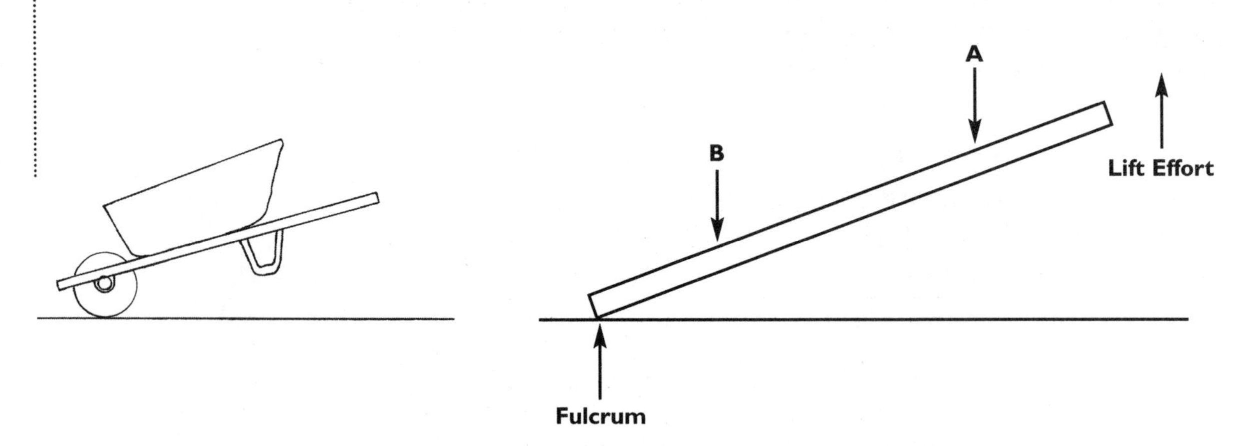

5. The placement of the fulcrum with the lever can be used to increase or decrease movement. As the fulcrum is moved each side of centre, there is a change in the amount of movement from one end of the lever to the other.

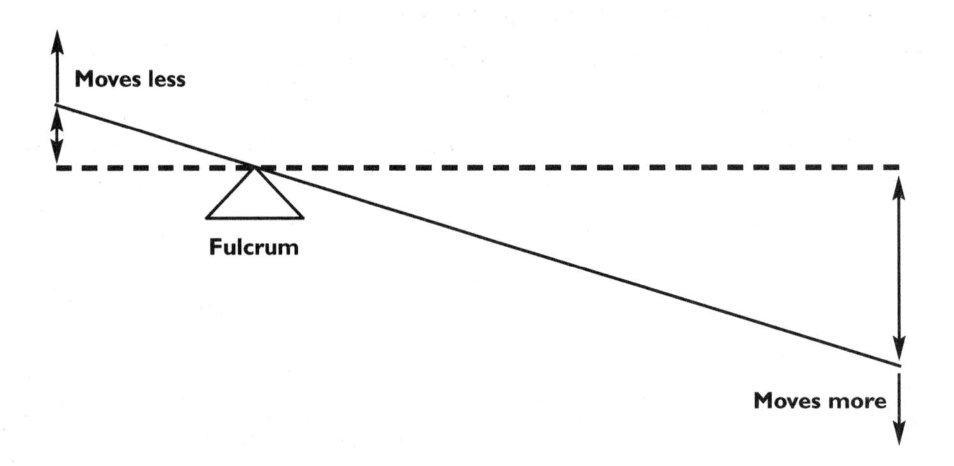

If hydraulic cylinders are used to raise the dump body on a truck, there is more movement required to raise the dump body when the hydraulic cylinder is furthest from the fulcrum (pivot), but this position also requires less effort. If the hydraulic cylinder was placed closer to the fulcrum, the reverse is true.

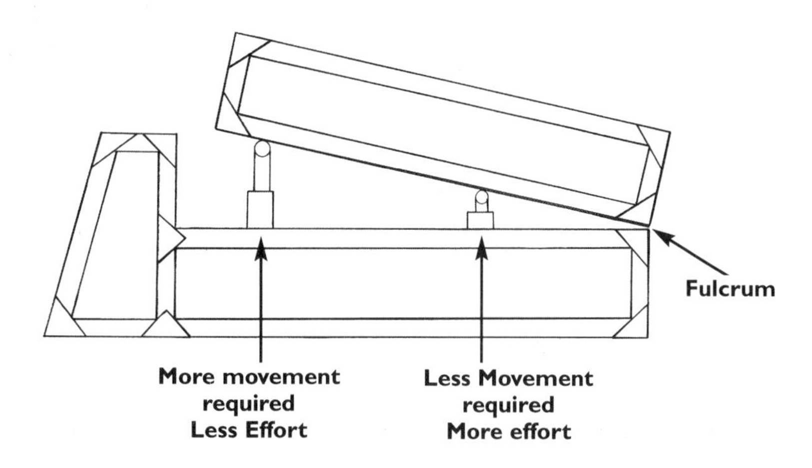

From the basic principles outlined above, students can use the lever principle to provide movement in toys and other devices. By using nothing more than card stock, art straws, and paper fasteners, the examples shown below can excite a lot of creative minds. Wings on birds can be made to flap, heads and tails on animals can be made to nod and wag, and arms and legs on people can be made to move. These patterns make excellent shadow puppets when a light beam is shone past them onto a screen.

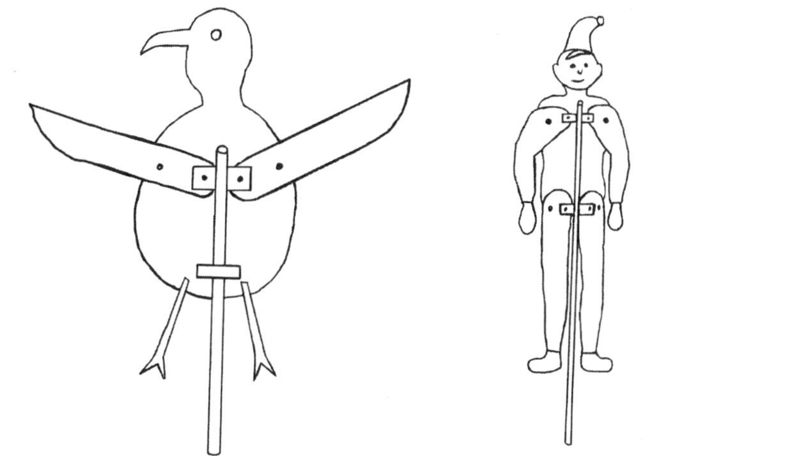

WHEELS AND AXLES

One of the most important inventions of all time is the wheel and axle. Wheels working together with axles make it easier and faster for us to move objects. Objects were probably first moved by using rolling logs as the earliest type of wheels. These rollers reduced the amount of effort caused by the friction involved in dragging objects across a surface. Students can see the difference in this effort by measuring the stretch of an elastic band attached to a mass. By pulling the mass with and without the use of pencils as rollers, the reduction in effort becomes obvious.

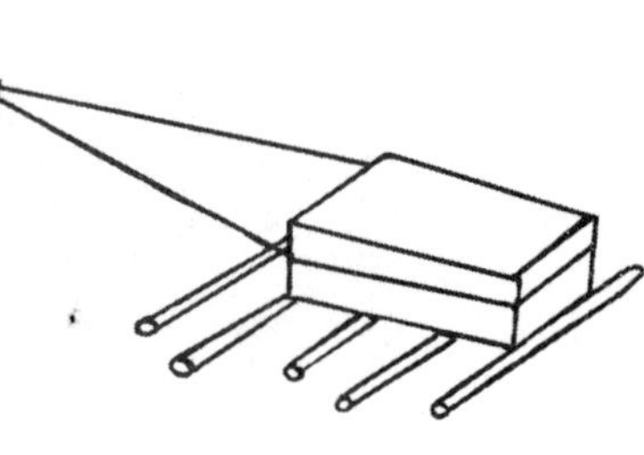

45° set square

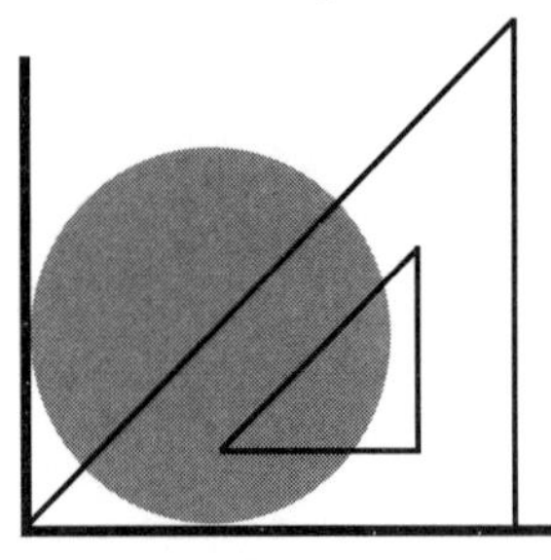

Rotate the lid to mark centre lines crossing one another

Although many applications for the wheel are possible, its use on student-designed vehicles is the most common. Wheels can be made from jar lids, lids from plastic containers, thread spools or pop cans, but purchased wooden and plastic wheels are the easiest to use when attaching wheels to axles. Axles can be made from wire rods, barbecue skewers, or nails, but the 5 mm dowel rod that fits the pre-drilled holes in purchased wheels is easier to use. When using lids and pop cans for wheels, care must be taken to locate the centre line for the axle accurately. Placing the lid or can in a corner of a bookshelf or other square corner and using a 45 degree set square as shown left, is an easy method in finding the exact centre.

The method by which the wheel is fastened to the axle offers students some choice. If the wheel is to rotate freely on a rigid axle, the wheel hole may need to be enlarged slightly to make this rotation frictionless. The wheel can be held in place on the axle with the use of short lengths of plastic tubing placed on each side of the wheel on the axle or rubber tap washes can be used in place of the tubing. If the wheels are to be secured to the axle, the axle has to rotate freely in a "bearing." Wood dowels fit nicely in jumbo Art Straws that make an excellent bearing. Alternatively, dowel axles can rotate in holes punched in card stock gussets that are in turn secured to wood strip frames as shown below.

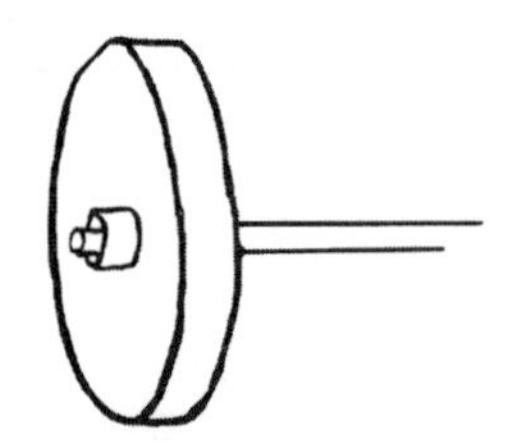

Wheels rotate on the axle. Pieces of tubing hold the wheel on the axle.

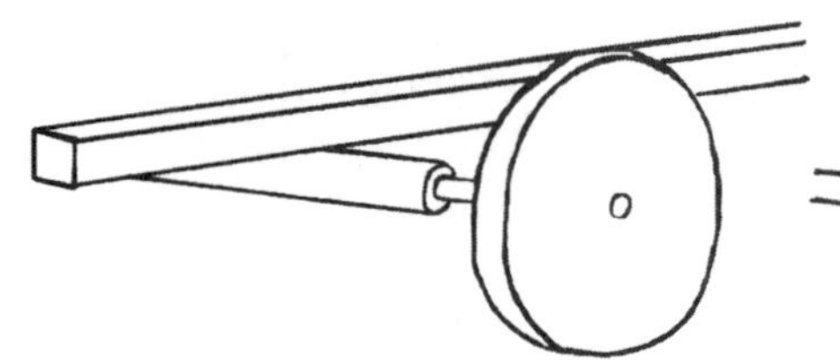

Wheel is secured to the axle that rotates in an Art Straw fastened to the vehicle frame

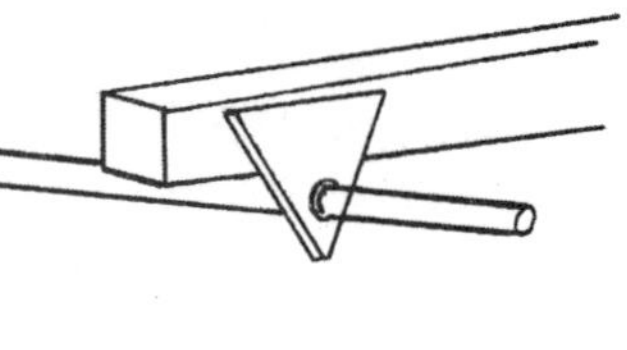

Axle rotates in a hole punched in a gusset.

PULLEYS

The pulley is another type of wheel that has a groove in its edge to guide ropes, cables, or belts when used with machinery. Pulleys help reduce friction while helping to lift objects. Uses of the pulley ranges from the raising of a flag on a flagpole, to the lifting of heavy objects with a building crane. Pulleys, linked by a belt, are used to transfer movement from one part of a machine to another. When pulleys of different sizes are linked, a change in rotation speed results.

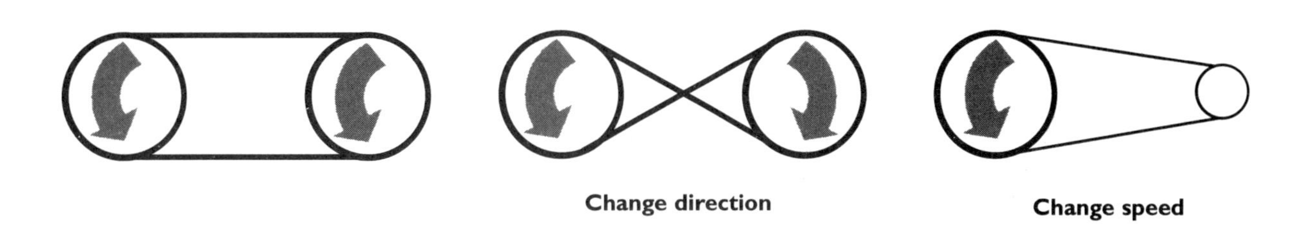

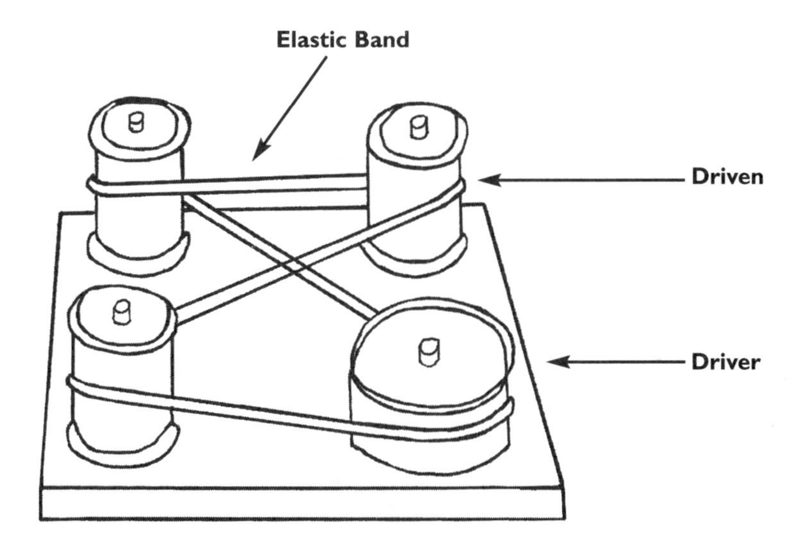

Have students make a gear board similar to the one shown above. Link a combination of thread spools and small cans of larger diameter with an elastic band. Notice the direction and speed of rotation of the driver and driven pulleys.

For students who have achieved a degree of skill in adding control to their devices, the pulley offers a wide range of design possibilities. Plastic pulleys of various diameters are available and with their pre-drilled holes, fit onto wood dowel rods that are used as axles or shafts. Elastic bands can be used as the belt to link pulleys or string or butcher's cord can be used as the cables for lifting devices.

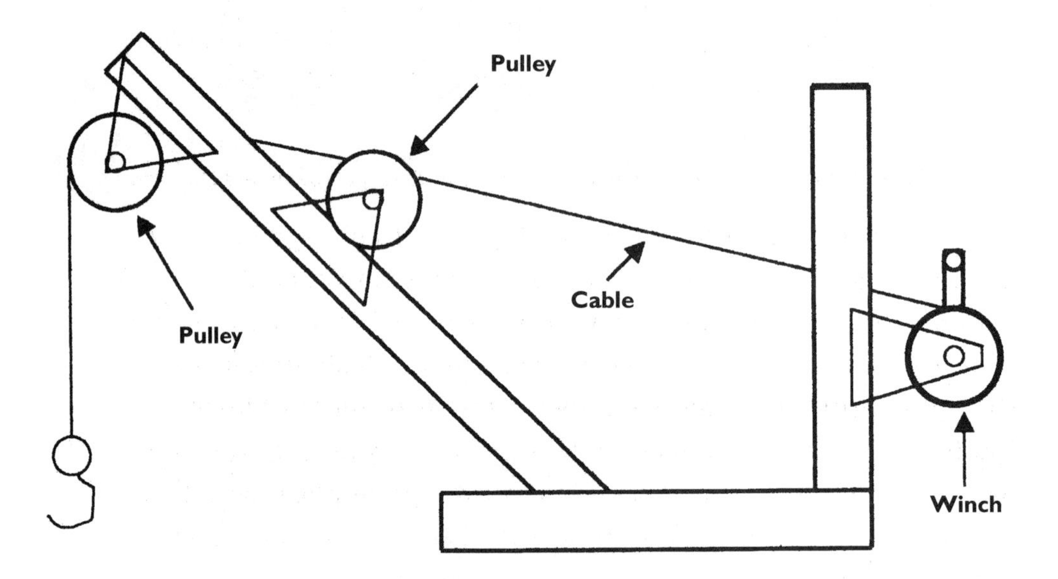

GEARS

Like the pulley, the gear is another type of wheel that transfers movement from one part of a machine to another. Several types and sizes of plastic gears are available for classroom challenges with the spur, bevel, and worm gear, being the most popular. Gears have teeth that mesh with one another and, unlike pulleys that are linked some distance from one another by a belt, gears must touch each other.

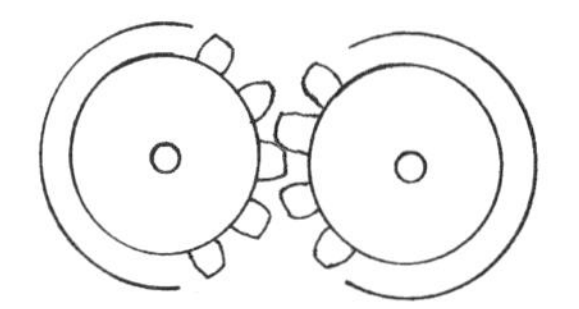

Typical Gear Mesh (Spur Gear)

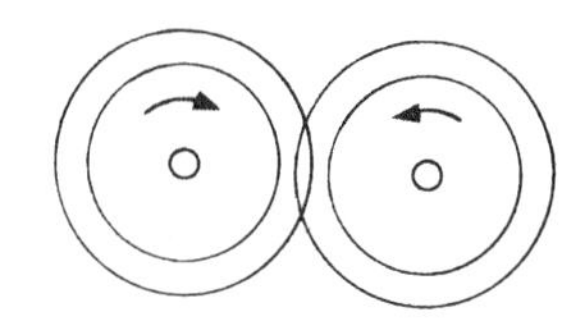

Change Direction

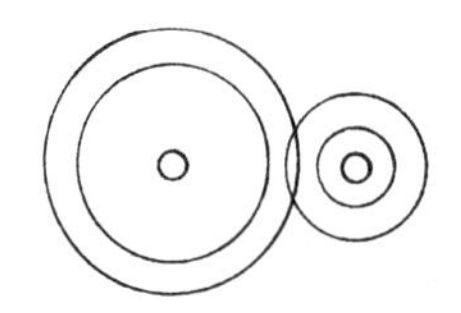

Change Speed and Direction

Several gears are often grouped together to change direction or to change their rotation speed. The cluster of gears within an analog wristwatch, or the larger Grandfather's Clock, are examples of this. For classroom use, plastic gears are available in various diameters and are usually described by their number of teeth. **Students should be encouraged to experiment with gear direction and rotation before using these in their design applications.** They will find that turning a 32-tooth gear will cause a 16-tooth gear to make two full rotations for every single rotation of the larger gear. Similarly, turning a 16-tooth gear one full rotation will only turn the 32-tooth gear one half of a full rotation.

Using the bicycle, as an example, is an excellent way for students to understand the gear to speed relationship. They should also be able to make relationships with "effort" when considering bicycle gears. They will find that under load, the larger gear needs more effort to drive the smaller gear, whereas the smaller gear uses less effort to drive the larger gear. This relationship is clear with the 10-speed bicycle.

The diagrams below show some additional gear types. The Worm gear is usually mounted on the shaft of a motor and drives a Spur gear when a 90-degree change in gear direction is required. The Rack gear is used when linear (straight line) movement is required. Bevel gears also allow for a 90-degree change in direction.

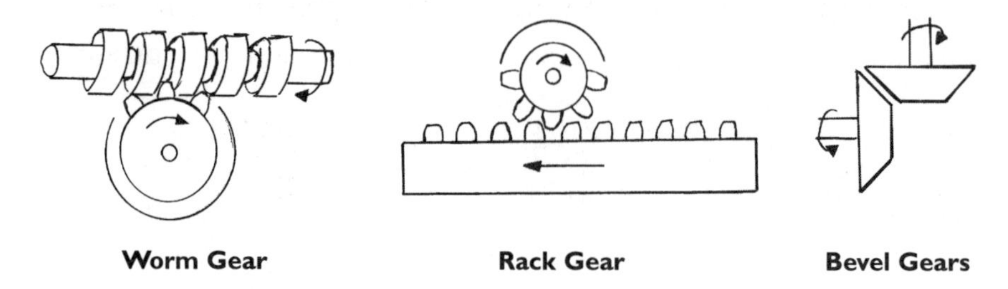

Worm Gear **Rack Gear** **Bevel Gears**

Gear Jigs are available for students to make their own gears that are suitable for some applications. Short lengths of dowel are sandwiched between two card stock or wood discs.

Younger students might try making gears by cutting "V" notches in the raised outer rim of paper plates. One or more sizes of these plates can be meshed together by placing a thumbtack, through their centres, onto a flat surface. Strips of corrugated cardboard (with the corrugations facing out) can also be glued around thread spools to make gears. The tooth-like edges on bottle caps, when placed close together, can also be used to show the meshing of gear teeth.

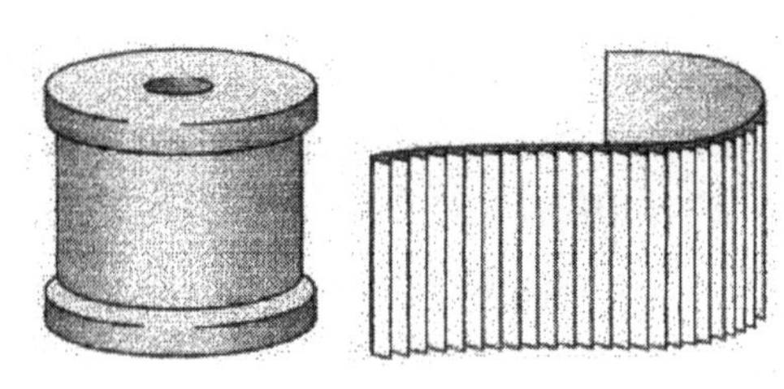

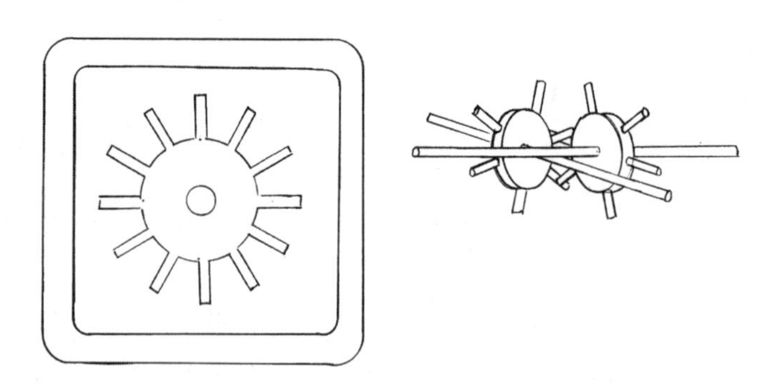

CAMS

When turning (rotating) movement is required to transfer movement to an up and down (reciprocating) motion, a Cam is used. Cams can be oblong (pear shape) or circular. They differ from the symmetrical rotation of a wheel on an axle by having their axle centre-line closer to their edges.

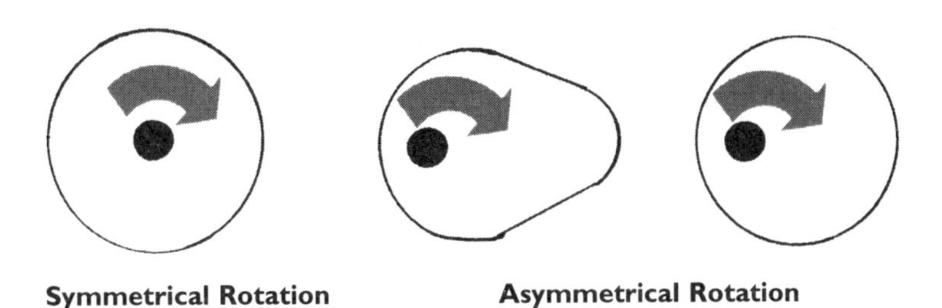

A cam is usually "fixed" (secured) on its axle and as the axle and cam rotate together, it transfers its asymmetrical up and down motion to a "follower." The follower, in turn, moves through a guide to provide straight-line vertical or horizontal movement.

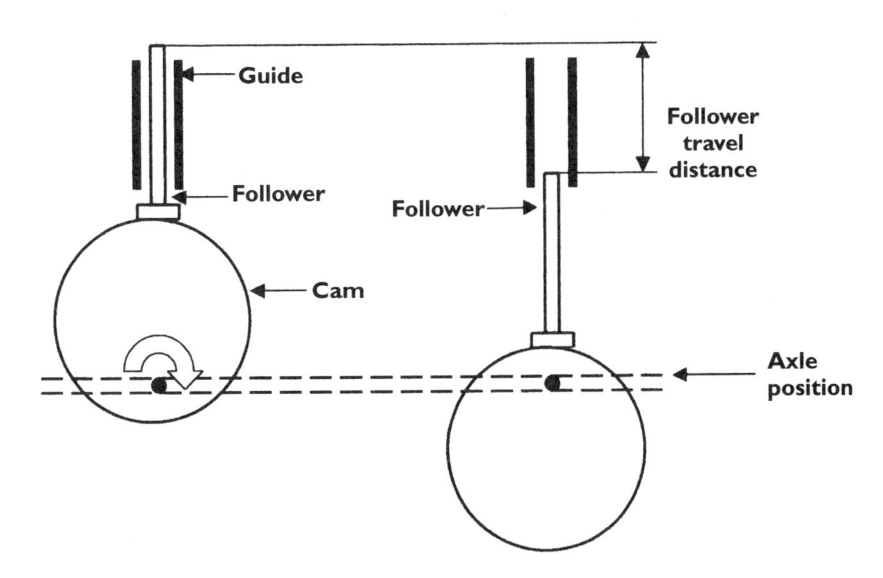

Students who are somewhat familiar with the automobile engine will recall that the "camshaft" that has a series of cams that link to valve stems (followers) that in turn open and close valves.

A cam can also be used to provide movement to a hinged lever as shown below. Have students explain what would happen to the lever if a cam of the same size was moved closer to the hinge.

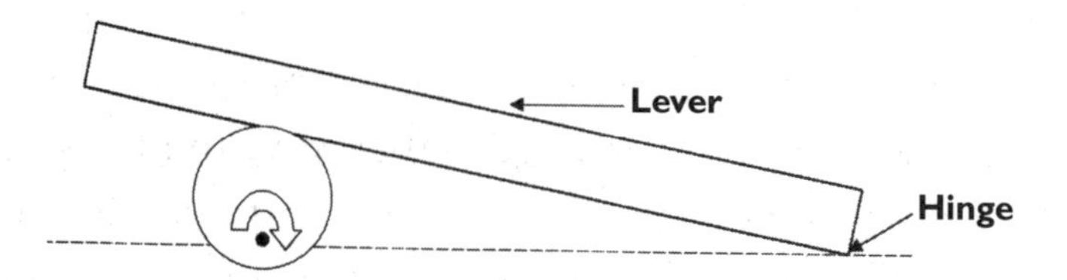

Drilling off-centre holes in wooden wheels can make excellent cams, or alternatively, plastic cams can be obtained from suppliers. Cams and followers can be assembled as shown below. The axle can be turned manually with a crank at one end of the axle, be driven by an electric motor linked to a pulley on the axle, or be driven by wheels that are part of a vehicle pulled along the floor. The latter method provides students with some creative and entertaining movement of objects and figurines. Try getting some mobile animations by using one or two cams on both axles of a pull-toy.

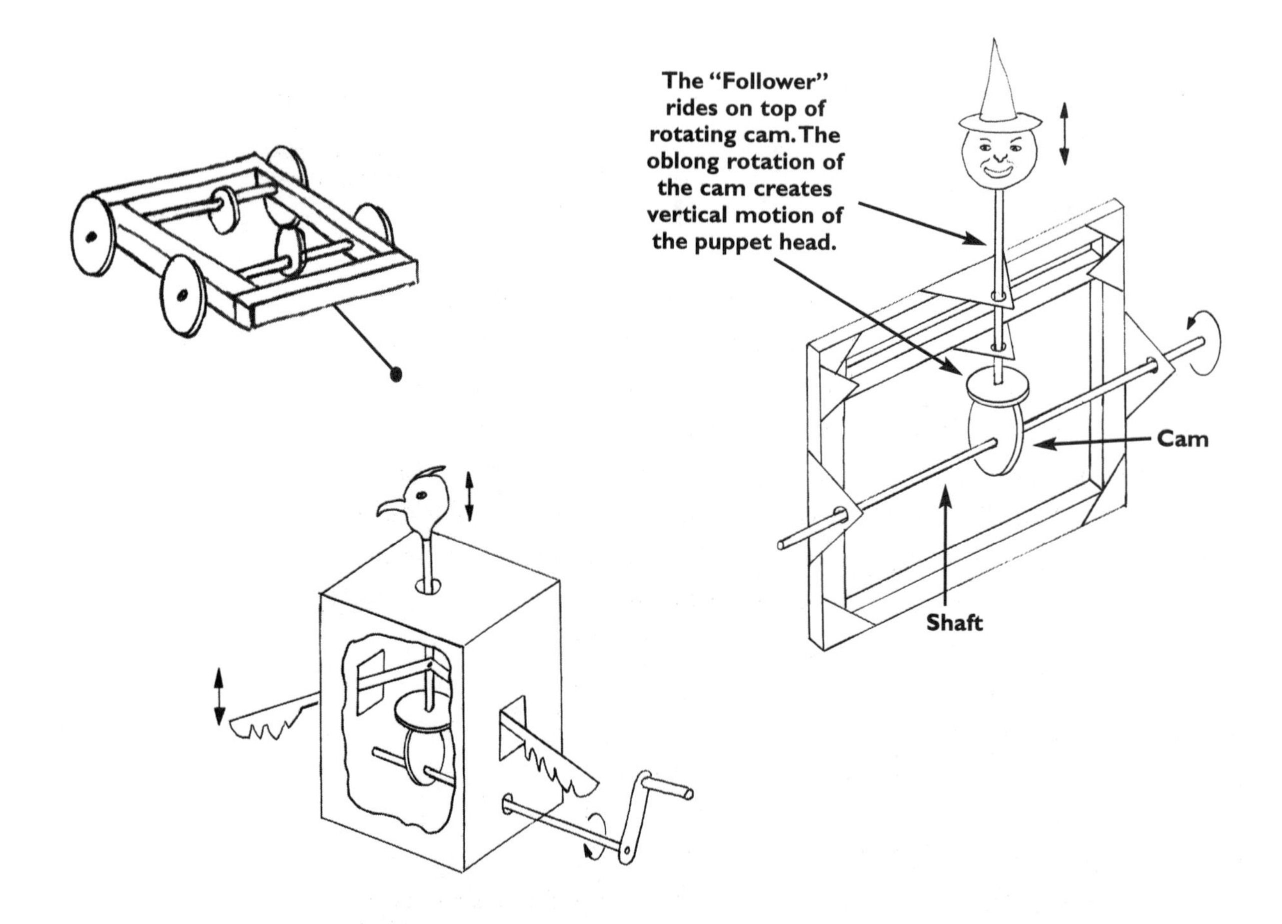

WINCHES

When loads are to be lifted or moved in some direction, a "winch" will provide the mechanical advantage (efficiency) to lift the load with ease. Winding a cable around an axle or drum provides this lift or movement. We can see examples of a winch in some of the old water wells that had to bring the water to the surface in a wooden bucket. Drawbridges on castles were raised and lowered with a winch. Boats are hoisted onto their trailers with a simple winch system.

Whether winches are operated with a crank handle as shown below, or driven by a motor, the "effort" required to turn the axle is determined by the length of the crank handle or pulley diameter that drives the axle. The longer the crank handle or the larger the pulley diameter used to turn the axle, the less the effort required.

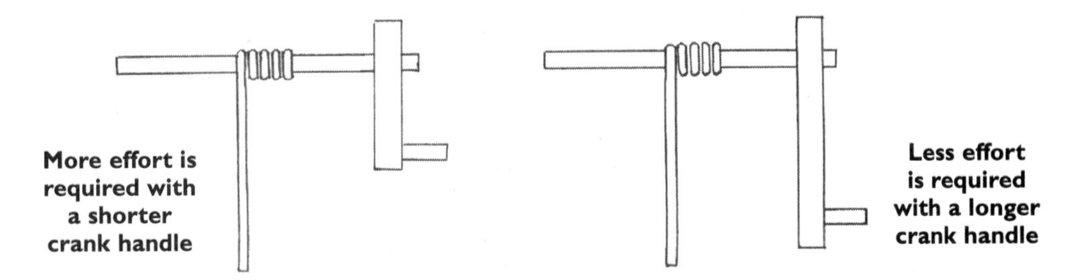

The amount of effort required to lift a load is also affected by the diameter of the surface onto which the cable is winding. If the axle is driven by the same crank handle length or pulley diameter, the effort is less when the cable is winding around a smaller diameter. Although this smaller diameter gives less effort, it also has less "speed" in lifting the load. The larger the diameter around which the cable is winding, the greater the speed of the lift. Often, a compromise between speed and effort has to be made.

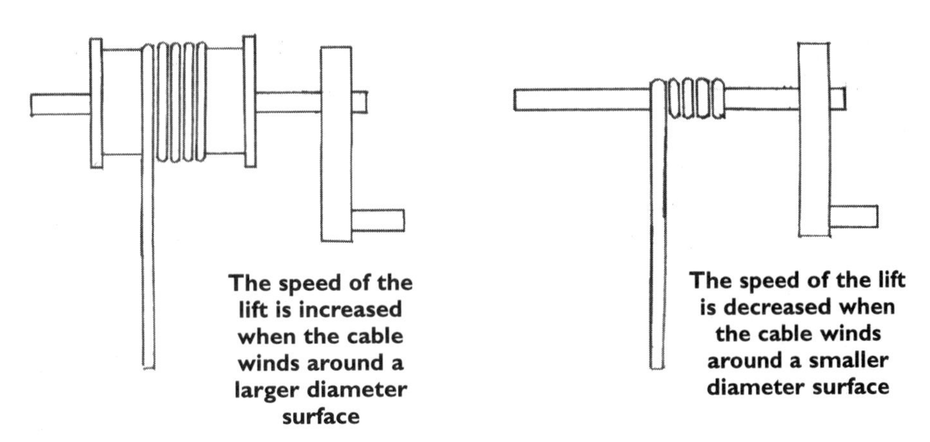

HINGES

A hinge is a "fixed" fulcrum around which levers pivot. The most common use of a hinge is on doors in our homes, but many machines require levers to pivot at a hinged location. Earlier when discussing levers, it was pointed out that the amount of effort required to activate the lever, depended on where the load (effort) was applied. The further the effort is from the hinge (fulcrum), the lesser the effort required. Have students think about the door to the classroom. Why is the door handle always placed furthest from the hinge?

Whenever a hinge is required for student challenges, there are several ways, as shown below, to make this possible without having to purchase these as hardware items. Holes drilled in card stock gussets can be used to hinge frame sections. Holes drilled through wood strips make an easy-to-assemble pin hinge. Art Straws glued to the edge of cardboard pieces make a hinge that closely resembles a door hinge. The flaps of this hinge can be glued to individual sections of a structure. Each of the hinge examples are joined with a dowel rod used as the hinge pin. Hinges can also be made with bookbinding tape, scoring a hinge line on cardstock, or cutting through all but the outer layer of plastic wafer board (Coroplast).

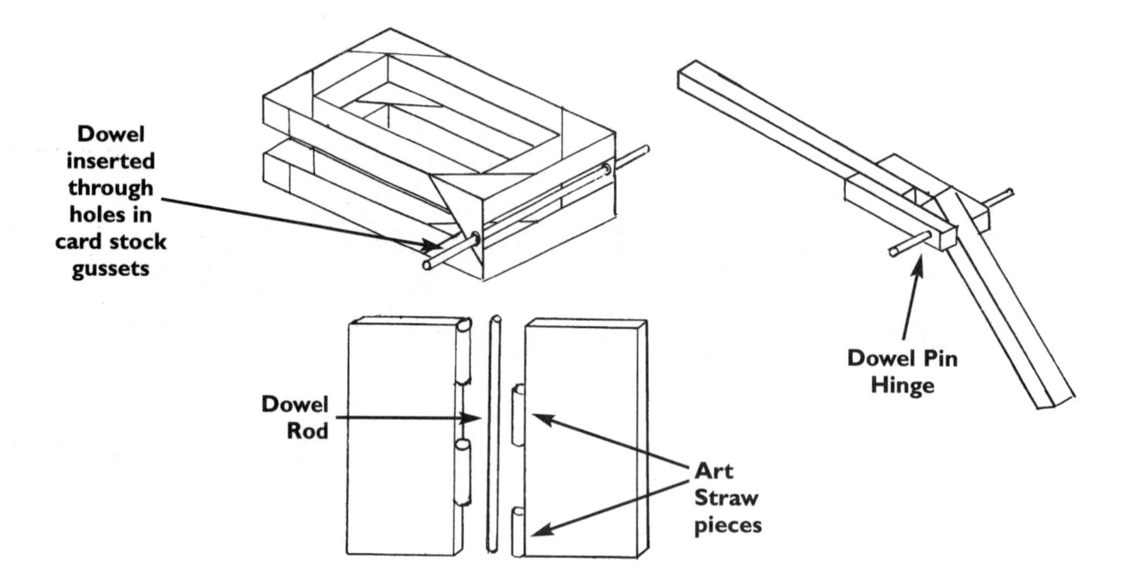

Section 9

Energy and Control Systems

Without energy, the structures, machines, and mechanisms that students have made would be lifeless. The two forms of natural energy that have been harnessed and used effectively are wind and water. Students may wish to design and build mechanisms that include a waterwheel or windmill. They could also make vehicles and boats, both with sails, and move these by providing wind energy from a fan. Miniature solar panels can also use the sun or artificial light sources to operate these machines and mechanisms.

MECHANICAL ENERGY

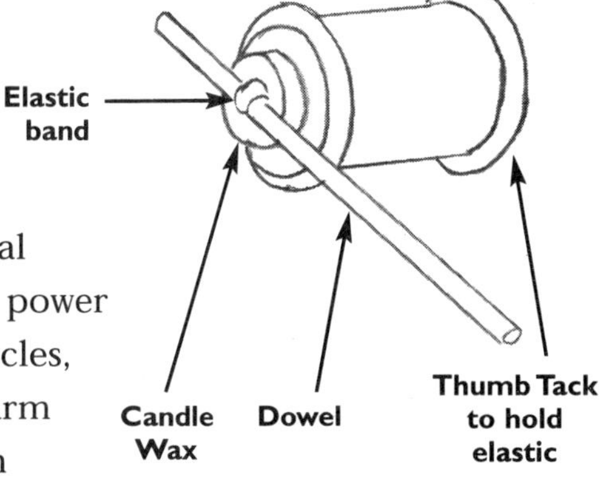

Elastic Bands

The energy stored in a stretched elastic band (potential energy) can be used in several applications to provide power to devices. The elastic band can provide power to vehicles, to provide motion to a mini-tank, and to activate the arm of a catapult. With each, students can experiment with different lengths and elastic band widths to see if one provides more stored energy than another.

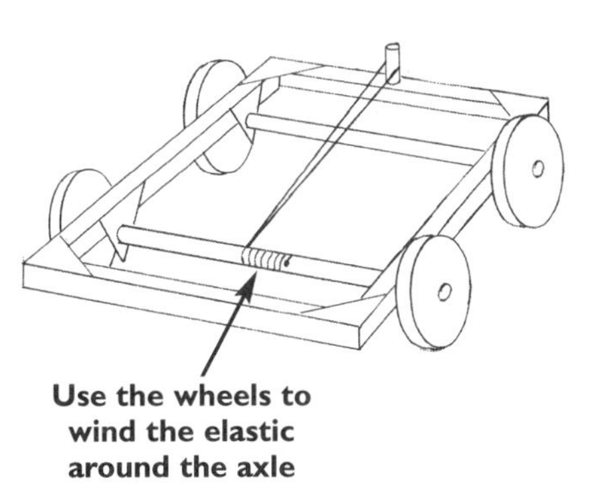

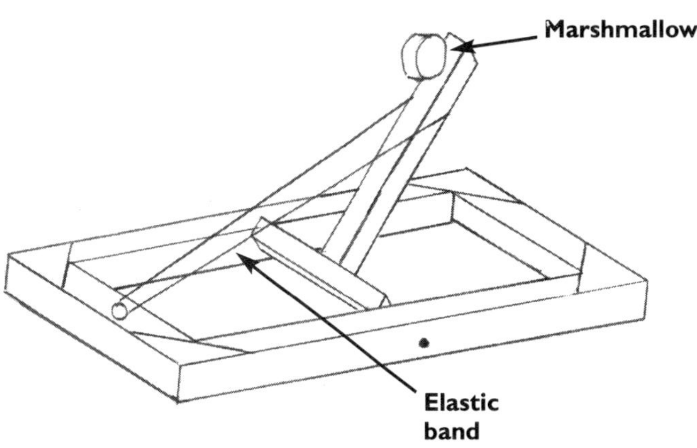

Syringes and Tubing

Pneumatic and Hydraulic systems are easily duplicated using syringes and tubing. These systems can be used to power mechanisms from a remote location. Syringes and tubing can be used with or without water in the system, but since air is easily compressed (pneumatics), the system is not as effective as hydraulics. Since liquids can not be easily compressed, filling the syringes and tubing with water (hydraulics), power is more effectively transferred to any mechanism.

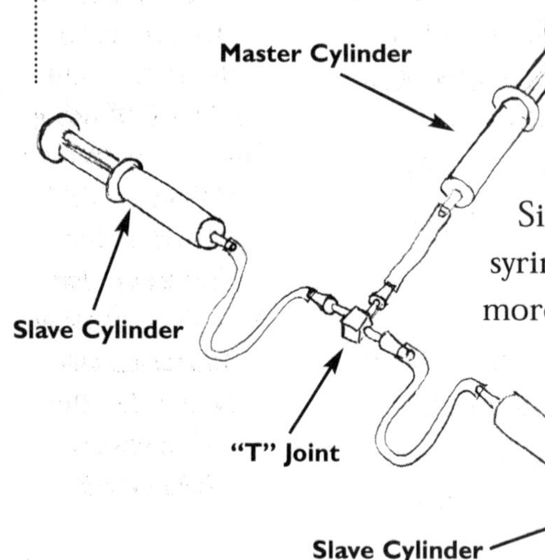

Have students fill a syringe with water and put their thumb over the nozzle end and attempt to compress the plunger. Have them remove the water and repeat the attempt to press the plunger. They will notice the difference in compressing liquid and air. Have them connect two different size syringes to the ends of a short length of tubing. They will notice a difference in the amount of plunger movement when alternately compressing the larger and smaller syringes. Does the amount of plunger movement differ on each end if the syringes are the same size? What happens to the cylinder plunger on the opposite end when you withdraw a compressed plunger?

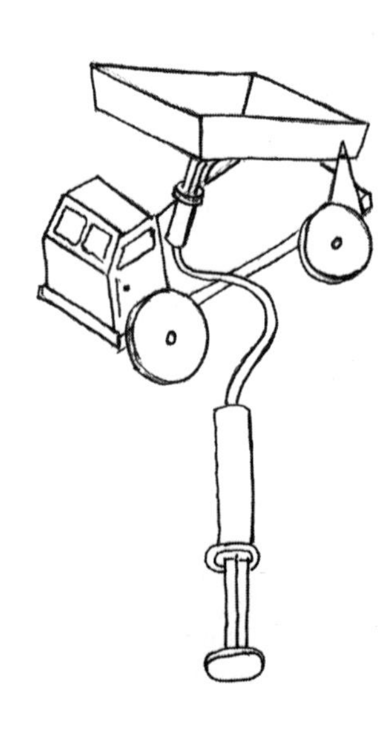

The syringe that is in the hands of the operator is called a "Master Cylinder." The syringe that transfers the power to the mechanism is called a "Slave Cylinder." With the use of a "T" or "Y" joint in the length of tubing, a single master cylinder can provide power to two slave cylinders.

Some of the applications shown are examples demonstrating control of a device. As students become skilled with their applications, control of mechanisms or devices from remote locations, are only limited by the imaginations of the designing student.

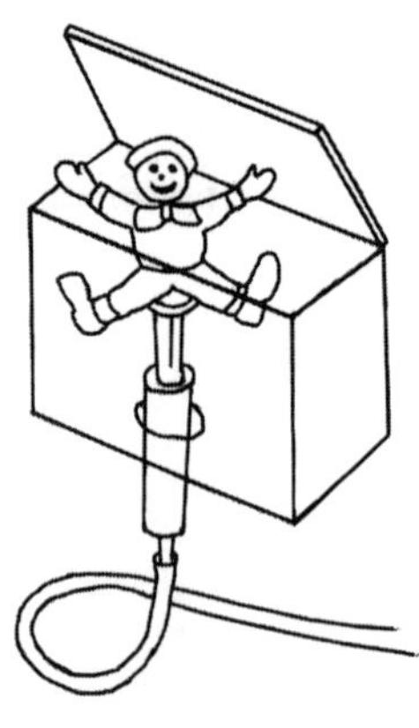

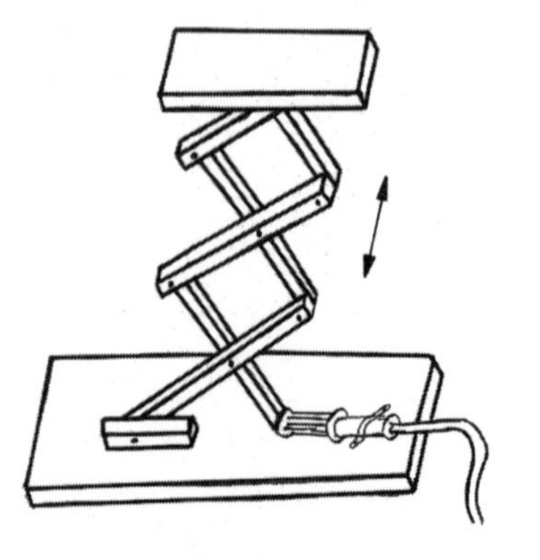

ELECTRICAL ENERGY

The electrical energy stored in batteries will provide the amount of power needed to run the electrical devices used in the challenges. Electrical adapters are also available that plug into wall outlets to convert to the low voltages required for some low voltage electrical devices. These are ideal for running devices like windmills or lifting devices that do not have to move from a stationary location. Batteries are better suited for vehicles that need to move about.

SAFETY FIRST!

Students should be regularly warned of the dangers of using the 120 volt power outlets in their home or the classroom for their investigations. Batteries will suffice for the challenges presented.

Component Voltages

The unit used to identify the "electrical force" of electrical devices used is voltage (v). This is important when matching the power source (battery, adapter) to the electrical component it is powering. A range of voltages available for the more commonly used electrical components is as follows:

BATTERIES (Cells)

Batteries are a series of "cells" connected together to produce a higher voltage.

- "D" and "C" size — 1.5 v
  (re-chargeable batteries should be considered)
- larger (square) lantern battery — 6 v

ADAPTERS (Voltage Converters):

- 120 v converted to 4.5 v or 9 v

MOTORS

Motors come in low, medium, and high-torque (a force that produces the strength of the rotation).
Low torque - 1.5 to 3 v
Medium torque - 2.5 to 4.5 v
High torque - 3 to 6 v

LIGHT BULBS (Lamps):

1.5, 2.5, and 3.5 v

BUZZERS

1.5, 3. and 6 v

Circuit Diagrams

For an electrical system to work, electrical current must complete a full circuit from the negative (–) terminal to the positive (+) terminal of a battery. To do this, it must pass through an electrical wire (conductor). If the current flow is interrupted (i.e. with an open switch) a full circuit is not completed. Switch positions can be "open" or "closed." If a light bulb is introduced in the system, evidence of the interruption or completion of the circuit is seen.

The following schematic is the conventional way of diagramming this circuit. It is often quicker and less confusing to use symbols instead of representative sketches of components to diagram each circuit. Students should be encouraged to diagram their circuits before putting them into practice, although you may need to provide considerable assistance with this step at first.

Circuit Diagram

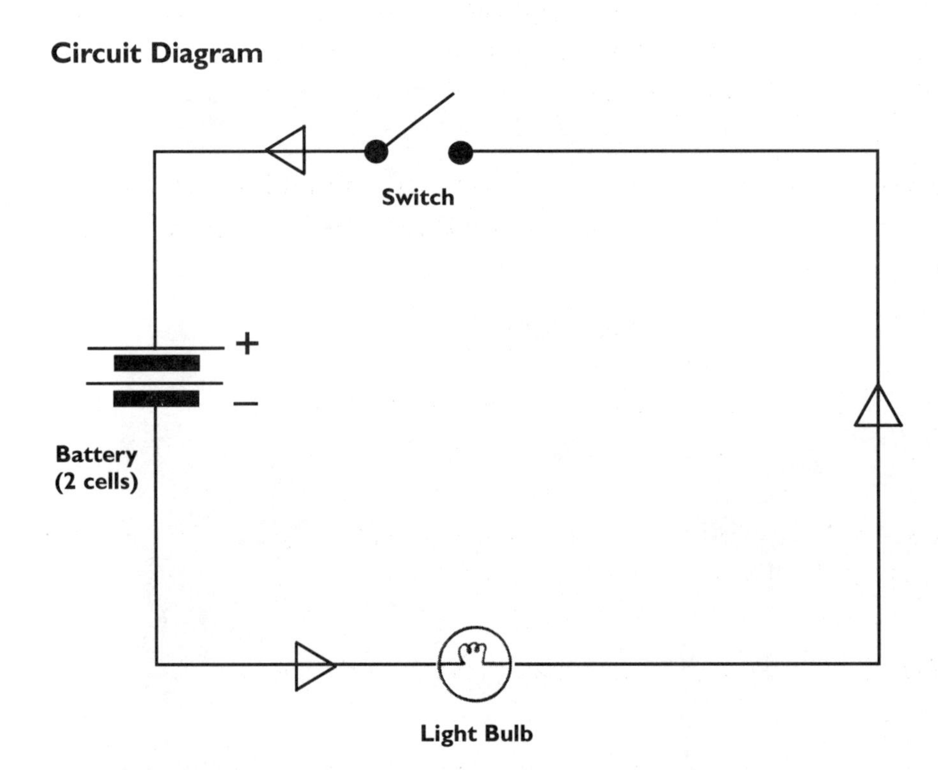

+ = positive terminal
– = negative terminal

Electrical Symbols

The following is a list of the symbols that are used to understand a circuit diagram.

Wire: As the previous circuit diagram shows, wire is sketched as a straight line even though in practice wire usually takes an irregular path from each electrical component.

Unattached Crossed Wire: Often, diagrams have to show one wire crossing another. This symbol is used when they have to cross but are not attached in the circuit.

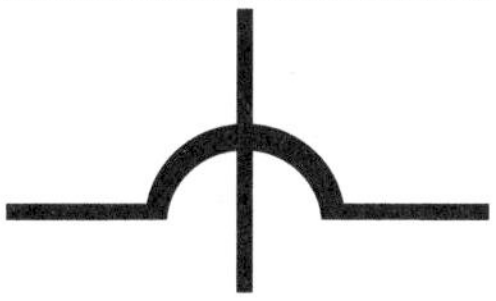

Attached Crossed Wire: This symbol is used whenever a wire is spliced (connected) to another wire.

Battery: For most electrical applications, this symbol indicates that batteries are used as the energy source (see the circuit diagram). Electrical Adapters may also be used as the energy source.

+ −

Switch: Whenever a light bulb is to turned off and on, or a motor stopped or started, these symbols indicate that a switch is needed in the circuit. For ease of understanding the diagram, the switch is always shown in its "open" position.

Open

Closed

Motor: This symbol is used to indicate that a motor is used in the circuit to drive a mechanism.

M

Buzzer: This symbol indicates that a signaling devise (buzzer) is part of the circuit.

Bell: This symbol indicates that a signaling device (bell) is part of the circuit

Light Bulb: This symbol indicates that a light bulb in the circuit is being used as a visual signaling device or as a headlamp in a vehicle's electrical circuitry.

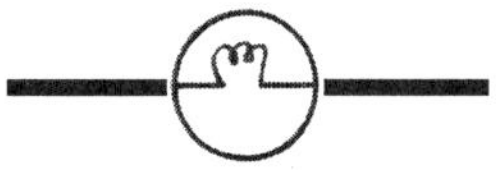

SAFETY FIRST!

Only use Paper Clip switches or variations of this switch with battery or adapter power sources. Use of 120 volt power sources in homes or classroom with these switches could result in serious injury.

Making a Simple Switch

There are several miniature switches available for the electrical control of devices, but these add unnecessary cost. The following switch can be made from available materials. Wire in the circuit is wrapped around two thumb-tacks and a paper clip is used to "open" and "close" the circuit. As long as contact is made between two metal surfaces connected by wires in a circuit, variations of this switch can be made. Aluminum foil is an excellent conductor and can be used in making switches. Foil can cover ends of wood strips and be slid together to close a circuit or brought together like a hinge. Students may be challenged to see how many variations of switches they can design and make.

Paper Clip Switch

Wire

Thumb Tacks

Paper Clip

Series and Parallel Circuits

Students need an opportunity to attempt a few circuits to get an understanding of a simple circuit. Having them try the following circuit is a good place to start. They should be able to trace (follow) the circuit from one side of the battery holder (– negative), through the switch and bulbs, and back to the (+ positive) terminal of the battery holder. This circuit is said to be in "series." Try putting another bulb in the circuit. Each bulb is sharing the electrical voltage but the brightness of each light bulb diminishes.

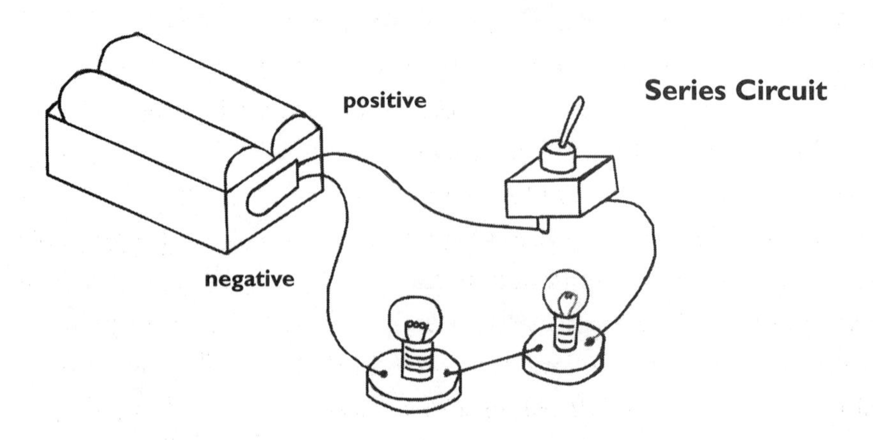

Series Circuit

The circuit shown below is different from the "series" circuit. The voltage flow still completes a circuit, but in this configuration, the circuit is said to be "parallel" and each light bulb will retain the same brightness as one by itself.

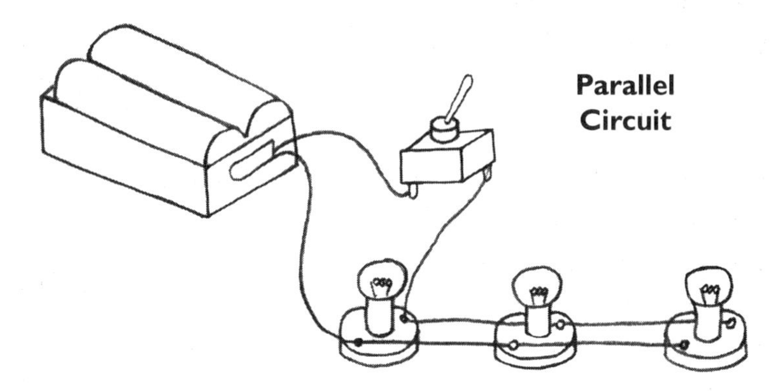

A circuit diagram for the Parallel circuit would look like the following:

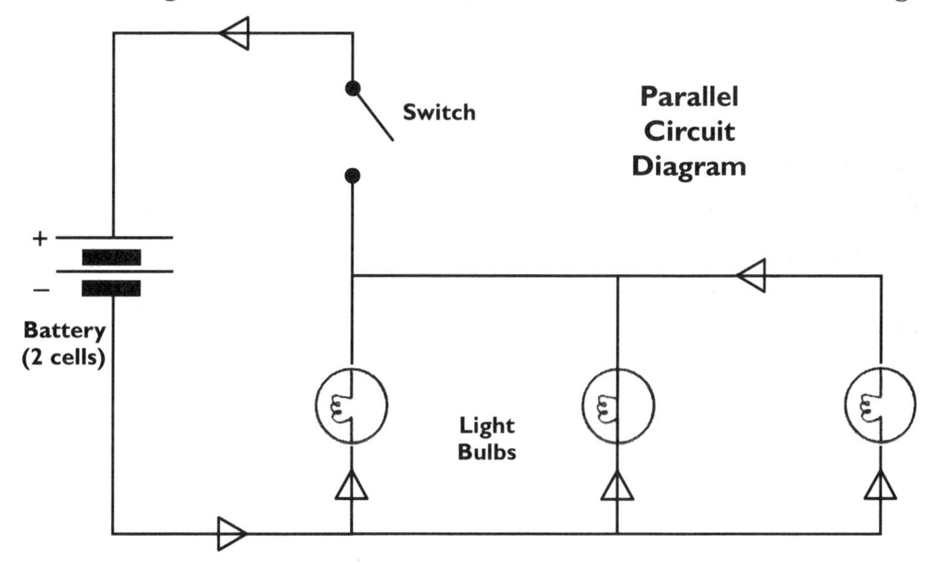

Fastening Electrical Components to Structures

Once students are comfortable with wiring a circuit using various components, the next challenge is finding ways to secure the components to their structures. A spot of hot glue will hold battery holders, bulb holders, and buzzers. Because the small electric motors require a secure location for them to link to gears and pulleys used to drive the mechanisms, mounting clips with their own sticky pad are available. Bulb holders are also available that slide on the ends of Popsicle sticks. The Popsicle sticks are easily glued to the frame of the structure.

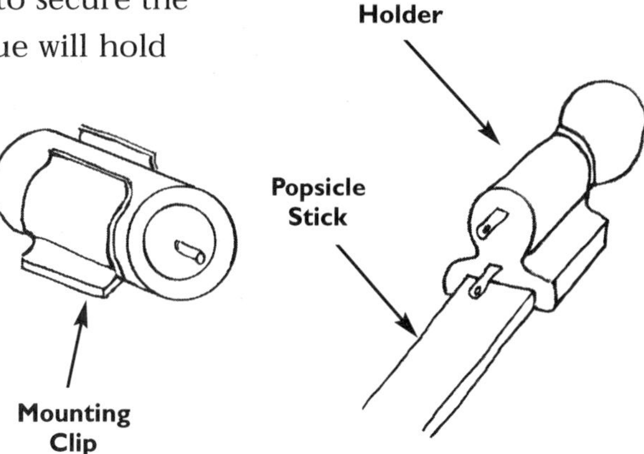

Control from a Remote Location

Very often, students will want to operate their device or mechanism from a short distance away. To accommodate this, lengths of wire can run to a paper clip switch mounted on a small piece of wood that can be held in their hands. Two lengths of wire are needed that run from the power source (battery) to the switch and from the switch to the device being operated. This control will allow an on/off feature to be added to control the operation of a vehicle.

Reversing the Direction of Motor Rotation

Sometimes, it may be necessary to reverse the direction of rotation of the motor. By reversing the wire from the battery to the terminals of the motor, the direction of motor rotation will be changed. An easier and quicker way to change direction while also maintaining the on/off feature is to modify the switch circuitry. The following circuit diagram shows how this can be done.

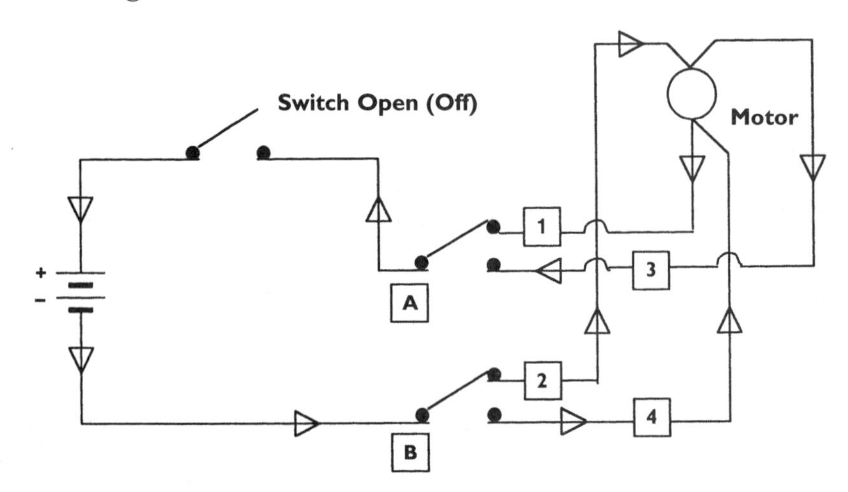

On/Off and Reversing Switch

- **Switches A & B closed on lines 1 & 2 for one direction of rotation.**
- **Switches A & B closed on lines 3 & 4 for the opposite direction of rotation.**

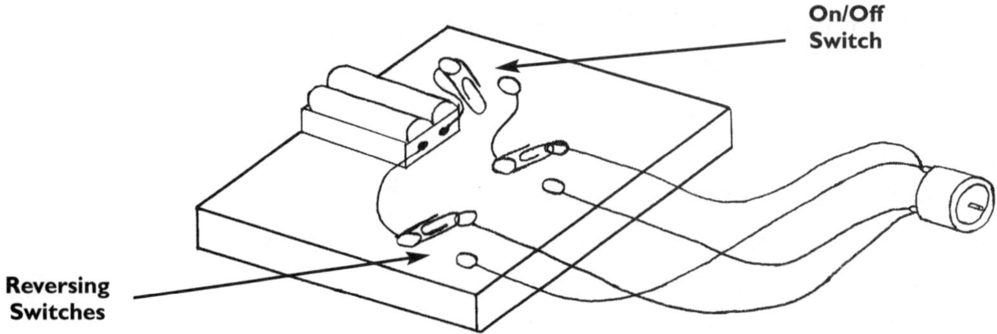

Section 10

APPROACHES TO ASSESSMENT

Before attempting the challenges in the next section of this resource, we need to consider some approaches to assessing learning as students attempt to find solutions to these challenges. It's not uncommon for teachers to think only of the **products** of their efforts, but as students experience opportunities to complete these challenges, we must assess the **process** of their learning experiences as well. Whether they are conducting experiments, making representative simulations, or assembling a product, we need to find ways to assess the acquisition of a new set of skills. By observing the process of learning, we will need to assess the development of skills that include creative thinking, problem-solving, and the development of appropriate attitudes.

Assumptions

Before developing assessment strategies, some basic assumptions need to be considered:

1. Although strategies for assessing student challenges are necessary elements in the teaching-learning process, **there is no single assessment vehicle that will provide a complete profile of the learner.**
2. Assessment must be learner focused. Whatever assessment strategies we generate, they must have the ability to assess students **as they learn, how they learn,** as well as **what they learn.**
3. Learning is an integrating experience. Throughout the process of learning, you must also take every opportunity to **point to the connections with other parts of the student's curriculum.**

Choosing an Assessment Strategy

It will be advantageous for us to choose one assessment strategy, or components of several strategies that work best for the student challenge. Factors contributing to our decision will include:

- the nature of the challenge being assessed;
- whether or not the strategy allows an assessment of a broad range of skills and behaviors;
- the time it will take to conduct the chosen strategy.

CHECKLISTS

Throughout the assessment process, we will be aware that students are supplying us with evidences of their developmental process, thoughts, and actions. It is essential for students to be aware that we will be observing these and that they will be supplying us with these evidences. As we ask students to account for their reasoning or verbal descriptions of events that are happening, we will need to capture these imaginative thoughts and actions in some way. This recognizes students as active participants and in many instances can be a motivation for them to improve these evidences.

The use of a Checklist with a set of "prompting" questions can be designed to encourage students to give us the evidences of the judgments they are making about their own observations, understandings, and actions. Types of prompting questions could include:

1. **Descriptive word lists:** sets of word prompts where students make judgments about their observations;
2. **Teacher/student interactions:** questions we can ask students to gain insight into their understanding and involvement with the challenge;
3. **Teacher observation:** questions we can ask ourselves about our perceptions of the understanding and involvement of each student with the challenge.

1. Descriptive Word Lists

The "Prompting" questions offered as examples below are suitable for early grades and for a challenge involving judgment of materials used in the challenge. Word lists will have to be tailored to the grade and components of the challenge.

Although such word lists can help us to assess a student's judgments about the materials they are using, lists can also assist in making connections to other parts of the curriculum. In this example, the checklist can be used to increase word power or to build a vocabulary of terms. Over time, students should be encouraged to move beyond simple yes/no answers to responses that are closer to complete sentences.

Word Lists

Challenge ______________________ ***Name*** ______________________

Are the materials you are using:

bright	colourful	glossy	dull
smooth	fuzzy	coarse	rough
soft	fluffy	hard	brittle
impressive	neat	attractive	ugly
strong	durable	flexible	weak
light	puffy	bulky	heavy
straight	curved	bent	twisted
massive	thick	slender	thin
huge	large	small	tiny

Summary Comments

____

____

____

____

____

____

2. Teacher/Student Interactions

The following is a representative sample of prompting questions that we can ask students to help us assess the **process of learning** as it is taking place and some judgments about the **product of their achievements.** Since it is impossible to observe everything that is happening in an activity-based environment, we must focus our observations to a manageable set.

Challenge ______________________ ***Name*** ________________________

Process — Specific Comments

Did you understand the challenge? ____________________________

What ideas did you contribute?______________________________

What did others think of your ideas? __________________________

Why was this idea/solution chosen? ___________________________

What procedure/materials did you choose to use? _______________

Why? ____

What part of the challenge will/did you work on? ______________

Did you help others with their part in the challenge? ____________

Did you have any difficulties with your part in the challenge?

____

Product

Did your product solve what you set out to achieve in the challenge?

____

Were there other materials/procedures you might have used? ______

Was your product safe to use? _________________________________

Can your product be used, or modified to be used, for anything else?

____

Summary Comments

____

____

____

____

3. Teacher Observations

The following is a set of questions we can ask ourselves as passive observers or through interactions with our students.

Challenge ________________________ ***Name*** ________________________

Teacher Observations **Specific Comments**

Does the student:

- have the ability to describe the challenge? ____________________
- take part in group discussions? ____________________
- share ideas and information? ____________________
- make suggestions? ____________________
- listen to others suggestions? ____________________
- participate well in the activity? ____________________
- cooperate well with others? ____________________
- show enthusiasm and motivation for the challenge? ____________________
- experiment with different materials? ____________________
- use tools and materials appropriately? ____________________

Summary Comments

RATING SCALES
(RUBRICS)

Where Checklists make either/or judgments about whether or not a student is moving toward a standard of achievement, Rating Scales make judgments about the intensity (how good?, how often?) of the achievement. Rating Scales are particularly helpful in assessing behaviors; what we see a student do or hear the student say. They can also be of value in assessing a particular skill, performance, procedure, ability, or product.

Rating Scales assume all students will exhibit skills and behavior in varying amounts, or at different rates, to a predetermined observation criteria. Rating Scales can be applied not only to the process and product of an activity-rich environment, but also to other areas of the curriculum.

Rating Scales with **descriptive categories** not only give us insights into the "process" the student is using, but also gives us evidences where support or remediation may be appropriate. Achievements on the Rating Scales may also provide evidence that the student can be challenged beyond the chosen criteria.

Teachers should develop descriptors that truly reflect the evidences that would be appropriate to the specific activity or challenge. Similarly, teachers should use an assessment criteria that reflects a continuum of performances or behaviors. For students to understand the meaning of each of the assessment criteria used, an explanation is required.

In the pages that follow, sample Rating Scales are provided with descriptive categories, assessment criteria, and explanations of assessment criteria you can use as examples to design your own scales. Some of the examples shown are representative examples of assessment and achievement projects presently used by educators or recommended by various guidelines.

Rating Scales with Descriptive Categories: Assesses the degree of achievement along an assessment criteria scale.

Process of Learning

Challenge ______________________ ***Name*** ________________________

1. Has the ability to describe the challenge.
   N/A Not at all In a small way To some extent To a great extent
2. Takes part in group discussions.
   N/A Never Seldom Sometimes Often Always
3. Shares ideas and information.
   N/A Never Seldom Sometimes Often Always
4. Makes suggestions.
   N/A Never Seldom Sometimes Often Always
5. Listens to others suggestions.
   N/A Never Seldom Sometimes Often Always
6. Participates well in the activity.
   N/A Never Seldom Sometimes Often Always
7. Cooperates well with others.
   N/A Never Seldom Sometimes Often Always
8. Shows enthusiasm and motivation for the challenge.
   N/A Not at all In a small way To some extent To a great extent
9. Uses equipment and materials appropriately.
   N/A Never Seldom Sometimes Often Always

Summary Comments

____

____

____

____

____

____

RATING SCALES WITH DESCRIPTORS OF THE ASSESSMENT CRITERIA

For students to understand the meaning of each of the assessment criteria used, an explanation is required. The first scale shows a generic template of a Rating Scale with each of the assessment criteria described for a stated student expectation. The second scale shows a representative sample of descriptive categories (expectations) and a brief explanation of each component of the assessment criteria used.

Generic Template

Assessment Criteria

Expectation	Beginning	Developing	Accomplished	Exemplary
Stated Expectation	Description of identifiable performance characteristics reflecting a beginning level of performance	Description of identifiable performance characteristics reflecting development and movement toward mastery of performance	Description of identifiable performance characteristics reflecting mastery of performance	Description of identifiable performance characteristics reflecting the highest level of performance

Rating Scale with Explanations of Assessment Criteria

	Beginning	Developing	Accomplished	Exemplary
		OR		
Expectation	Needs Improvement	Satisfactory	Good	Excellent
Participates well in the activity	Did less work than the others	Did almost as much work as the others	Did an equal share of the work	Did more than their share of the work
Listens to others suggestions	Is always talking and never allows others to speak	Does most of the talking and seldom allows others to speak	Listens, but sometimes talks too much	Listens and comments well to others ideas
Design Skills	Little to no layout and design skills	Simple design but layout could be more organized	Attractive design and layout that invites the viewer	Exceptional design and layout with outstanding visual appeal

Rating Scale with Assessment Criteria by Levels of Achievement

Many current curriculum guidelines have assessment and evaluation based on student expectations and assessment criteria based on levels of achievement. The following scales have the levels with a more complete explanation of the assessment criteria.

Assessment Criteria

Expectation	Level 1	Level 2	Level 3	Level 4
Applies the Design Process	• applies few of the required skills and strategies	• applies some of the required skills and strategies	• applies most of the required skills and strategies	• applies all of the required skills and strategies
Has the ability to describe the challenge	• does not demonstrate an understanding of the problem	• demonstrates a partial understanding of the problem	• demonstrates a basic understanding of the problem	• demonstrates a thorough understanding of the problem
Has the ability to develop a plan for designing the product	• no plan is attempted for designing a product, or the plan is incoherent or unworkable	• develops a plan for designing a product that is limited in appropriateness, efficiency, clarity, and completeness	• develops a plan for designing a product that is appropriate, clear and complete	• develops a reproducible plan for designing a product that is appropriate, efficient, clear and complete
Is able to follow the plan to build the product	• does not follow a plan to build a product	• follows most steps in a plan to build a product	• follows all steps in a plan to build a product, and makes required modifications	• follows all steps in a plan to build a product, and makes and records required modifications
Uses equipment and materials appropriately	• needs assistance to select appropriate materials and equipment to build a product	• selects appropriate materials and equipment to build a product	• selects appropriate materials and equipment and adapts materials to enhance the performance and design of the product	• selects appropriate materials and equipment to enhance the performance and design of the product
Modifies and re-tests the product	• makes no modifications or re-testing of the product	• makes modifications but does not re-test the product	• makes and records modifications and re-tests the product	• makes, records and justifies modifications, and re-tests the product
Product addresses the original need	• product does not address the original need	• product partly addresses the original need	• product addresses the original need	• product fully addresses the original need

Scored Rating Scale

The following Rating Scale uses numerical values to assess the "quality" of a student's performance or the product of the activity. This method asks us to select a numerical value that best describes that quality. When numerical values of 1 to 5 are used for example, we should attempt to design a statement that truly reflects the numerical value given to it.

Challenge ______________________ ***Name*** ________________________

Scale

0 = N/A – not enough information to judge or has not had an opportunity to apply the category.
1 = Needs improvement – exhibits minimal effort.
2 = Not up to expectations – has not achieved minimal expectations.
3 = Good effort – reasonable attempt to achieve expectations.
4 = Excellent effort – achieves all expectations.
5 = Outstanding effort – achieves beyond expectations and is a mentor to others.

Category

- Has the ability to describe the challenge.
- Takes part in group discussions.
- Shares ideas and information.
- Makes suggestions.
- Listens to others suggestions.
- Participates well in the activity.
- Cooperates well with others.
- Shows enthusiasm and motivation for the challenge.
- Experiments with different materials.
- Uses tools and materials appropriately.

Summary Comments

____

____

____

An Achievement Table

When numerical evaluations are required, the Ministry, Boards, and schools will likely provide directions with a summary description of achievement in each percentage and corresponding level of achievement. The following table is an example only.

Percentage Grade Range	Achievement Level	Summary Description
80–100%	Level 4	A very high to outstanding level of achievement. Achieves all, and in some instances, exceeds expectations.
70–79%	Level 3	A high level of achievement. A good effort is shown to achieve the expectations.
60–69%	Level 2	A moderate level of achievement. Exhibits a reasonable attempt to achieve the expectations.
50–59%	Level 1	A passable level of achievement. Exhibits a minimum of effort to achieve expectations.

Summary

When developing assessment strategies, we must consider the following:

- no single assessment strategy will give you a complete profile of the learner;
- assessment strategies must be learner focused;
- where possible, connections must be made to other parts of the curriculum;
- constructing Checklists and Rating Scales take time, but systematic observation of students can be a most effective tool for a comprehensive assessment as well as providing the foundation for discussions with our students, their parents, and other staff;
- developing descriptors, assessment criteria, and explanations of the criteria, will improve over time and with practice;
- select one or two types of skills or behaviors to observe at one time;
- limit observations to a few students at a time;
- assessing an activity-based program is not substantially different from assessment in other parts of the curriculum.

Section 11

ACTIVITIES AND CHALLENGES

It was mentioned earlier, that technological activities and challenges could arise from work already taking place in the classroom. Because many teachers will need help in identifying the stimuli or trigger for these technological activities or challenges, it is hoped the following suggestions will help to make this link. Each should be considered to be a problem-solving challenge and where possible, follow the design process. It should be remembered that students in the early grades, and those with little exposure to the design process, need a lot of teacher assistance not only in identifying appropriate technological activities, but also with suggestions to choose appropriate tools and materials and for assembly procedures.

The activity or challenge suggestions are presented in two ways. Some of the suggestions are offered as **"Teacher-directed Activities"** to consider what would enhance expectations of the curriculum, while others are presented as **"Open-ended Problem-Solving Challenges"** that would identify a need and a specific problem to be solved. The latter would be more suitable for the age group that is able to reason through and apply the design process.

For the most part, the following suggestions will stem from stimuli suggested by expectations within the Science and Technology curriculum and will make use of the processes, tools and materials outlined in this resource. It should be noted that although several student expectations precede each activity or challenge, there are several other expectations that could also be addressed. The abilities of students should be considered when choosing from the suggested activities.

TEACHER-DIRECTED ACTIVITIES

The following group of technological activities are specific challenges suggested to augment Science and Technology curriculum expectations. You will notice that challenges have not been suggested for all grades, for all curriculum strands, or for all student expectations. This does not mean that a challenge is not possible for these. Teachers are encouraged to develop their own challenges beyond the samples outlined.

With each, there is still some problem-solving and decision making to be made, but each is teacher-directed. With many of the challenges, suggestions are made to assist teachers, but students should be encouraged to come up with their own ideas. More open-ended and challenging activities are provided in the subsequent section.

A SUMMARY OF THE ACTIVITIES

Correlated by Strand, Topic and Grade to a sample curriculum

Strand	Topic	Grade Level	Activity
Life Systems	Characteristics and Needs of Living Things	1	• Design and make a device to simulate a frog's hop. • Design and make shadow puppets with moving arms, legs, and wings. • Design and make a pull toy. Make pop-up greeting cards.
Life Systems	Growth and Changes in Plants	3	• Design and make a container to grow plants • Design a method or device to keep plants wet during vacation time.
Life Systems	Habitats and Communities	4	• Make a spider's web. • Make a bird's nest or beaver dam. • Build a model habitat as a community of plants and animals. • Have students sketch and make a cardboard model of their own bedroom • Have students design and build an exercise area for a small pet. • Design, set up, and maintain a temporary aquarium habitat for small animals and plants.
Life Systems	Human Organ Systems	5	• Make a model of the shoulder, elbow, wrist, and finger joints. • Make a model of the skeletal structure of the body.
Life Systems	Interactions Within Ecosystems	7	• Design and build a mini "decomposer factory" (composter).
Matter & Materials	Magnetic and Charged Materials	3	• Design and make a sheet-board racetrack. • Make a device to test for magnetic/non-magnetic materials.
Matter & Materials	Properties of Air and Characteristics of Flight	6	• Make an aerofoil (wing section). • Make a hot-air balloon. • Make a parachute. • Make a hang glider. • Make a helioblade. • Make paper airplane.
Matter & Materials	Fluids	8	• Make a submarine. • Raise a sunken ship.
Energy & Control	Energy from Wind and Moving Air	2	• Make a kite. • Make a windmill.
Energy & Control	Forces and Movement	3	• Make an air-propelled box.
Energy & Control	Light and Sound Energy	4	• Make paper shadow images. • Make a periscope. • Make a stringed instrument. • Make a wind instrument. • Make a percussion instrument.
Structures & Mechanisms	Everyday Structures	1	• Make a bunk bed. • Make a ladder. • Make a fence and gate. • Make a wheelbarrow.
Structures & Mechanisms	Movement	2	• Make a junk-mobile.
Structures & Mechanisms	Stability	3	• Build a beam bridge. • Build a tower. • Make a lever puppet. • Build a crane.
Structures & Mechanisms	Pulleys and Gears	4	• Make a pulley board. • Make a gear board. • Make a crane.
Earth & Space Systems	Weather	5	• Build a weather vane. • Build an anemometer.
Earth & Space Systems	Space	6	• Build a water clock.

STRAND: Life Systems
TOPIC: Characteristics and Needs of Living Things
LEVEL: Grade 1

ANIMAL MOVES

STUDENT EXPECTATION

Describe the different ways that animals move to meet their needs.

Activity 1. **Design and make a device to simulate a frog's hop.**
(see sample design challenge on page 15).

Activity 2. **Design and make shadow puppets with moving arms, legs and wings.**
(see "Levers" in the Machines and Mechanisms section)

Activity 3. **Design and make a pull toy.** Use a "cam" to provide the mechanism for animal movements.
(see Cams in the Machines and Mechanisms section)

Activity 4. **Make pop-up greeting cards**
(with animals, birds, or insects).

STRAND: Life Systems
TOPIC: Growth and Changes in Plants
LEVEL: Grade 3

PLANT PLACE

STUDENT EXPECTATION

Describe the changes that plants undergo in a complete life cycle.

Activity 1. **Design and make a container to grow plants.**
You might consider using wood strips for the container frame. Get students to suggest materials that can be used to cover the frame. What can be used to make the container waterproof? Perhaps a decorative container could be designed to hold a plant pot.

Activity 2. **Design a method or device to keep plants wet during vacation time.**

- Plastic Bag covering plant?
- Plant Pot set in tray of water?
- Wick to transfer water from a glass?
- Can a device be made to drip water onto the plant?

STRAND: Life Systems
TOPIC: Habitats and Communities
LEVEL: Grade 4

BUILDING MODELS OF ANIMAL HOMES

STUDENT EXPECTATION

Recognize that animals and plants live in specific habitats because they are dependent on those habitats and have adapted to them.

Activity 1. **Make a spider's web.** Use thread or string. As an extension to this, you might have students make a variety of "Dream Catchers" to be hung in the classroom.

Activity 2. **Make a bird's nest or beaver dam.** Collect branches and twigs.

Activity 3. **Build a model habitat as a community of plants and animals.** What materials can you use to represent living and non-living things (rocks, lily pond, soil, foliage)?

Activity 4. **Have students sketch and make a cardboard model of their own bedroom.** Can the bedroom be redesigned to make it a better place to live?

Activity 5. **Have students design and build an exercise area for a small pet such as a hamster or bird.**

Activity 6. **Design, set up, and maintain an aquarium habitat to temporarily house small animals and plants from a local wetland.**

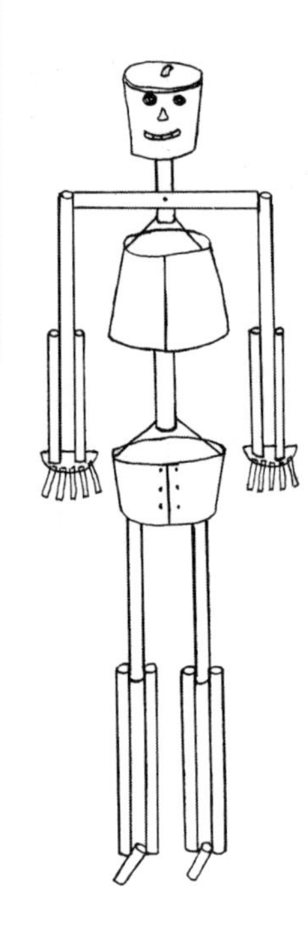

STRAND: Life Systems
TOPIC: Human Organ Systems
LEVEL: Grade 5

MODELING THE SKELETON AND JOINTS

STUDENT EXPECTATION

Describe, using models and simulations, ways in which the skeletal, muscular, and nervous systems work together to produce movement.

Activity 1. **Make a model of the shoulder, elbow, wrist, and finger joints.** Use paper tubes and elastic bands or wood strips that are hinged and activated by pneumatics (syringes and tubing).

Activity 2. **Make a model of the skeletal structure of the body.** Use paper tubes and rolls tied together with string.

INTERACTIONS WITHIN ECOSYSTEMS

STRAND: Life Systems
TOPIC: Interactions Within Ecosystems
LEVEL: Grade 7

STUDENT EXPECTATION

Explain the importance of micro-organisms in recycling organic matter.

Activity 1. **Design and build a mini "decomposer factory" (composter).** How big does it need to be? What materials can be used?

STRAND: Matter and Materials
TOPIC: Magnetic and Charged Materials
LEVEL: Grade 3

MAGNETS

Activity 1. **Design and make a sheet-board racetrack**

STUDENT EXPECTATION

Identify materials that can be placed between a magnet and an attracted object without diminishing the strength of the attraction.

How to Proceed: Students can guide a variety of materials to use as racers along the track with a bar magnet directed from beneath the sheet. What sheet materials can be used? How thick can the sheet material be? What materials can be used as racers? How fast can you make three laps and stay within the painted track?

Activity 2. **Make a device to test for magnetic/non-magnetic materials.**

STUDENT EXPECTATION

Classify materials that are magnetic and non-magnetic and identify materials that can be magnetized.

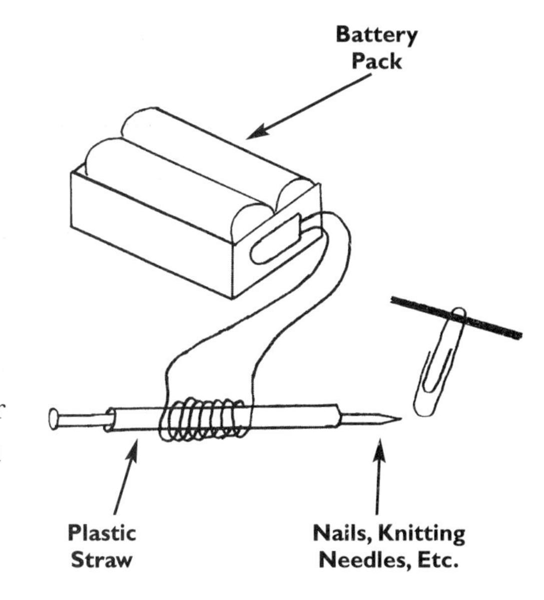

How to Proceed: Tightly wind 1 m of insulated wire around a plastic straw. Leave about 100 mm of wire on each end of the coil to fasten to the battery pack. Try inserting a variety of materials through the plastic drinking straw (steel and aluminum nails, short metal and plastic knitting needles etc.). To determine whether each is magnetic or non-magnetic, hang a paper clip on a piece of wood dowel or barbecue skewer and place the clip near the end of the nail or knitting needle. Is there any difference in the magnetism if the clip is placed on the opposite end of the nail or needle? What can be determined from this difference?

STRAND: Matter and Materials
TOPIC: Properties of Air and Characteristics of Flight
LEVEL: Grade 6

DESIGNS FOR FLIGHT

Activity 1. **Make an Aerofoil (wing section)**

STUDENT EXPECTATION

Demonstrate and explain how the shape of a surface over which air flows affects the role of lift (Bernoulli's Principle) in overcoming gravity.

How to Proceed: Fold a sheet of paper or card stock in half. Slide the top part of the sheet over the bottom piece 15 mm and glue or tape together as shown. The bottom of the wing should be flat and the top curved. Make a small tail (rudder) and glue or tape it in the centre at the trailing edge of the wing section. At the thickest part of the wing section and centred on its width, punch holes and glue a short piece of Art Straw at 90 degrees to the bottom of the wing. Feed 600 mm of thread through the straw. Holding each end of the thread, place the wing in front of a fan or hairdryer. The wing should work its way up the thread. Why does this happen?

Activity 2. **Make a Hot-air Balloon**

STUDENT EXPECTATIONS

Demonstrate that air expands when heated.
Design, construct, and test a structure that can fly.

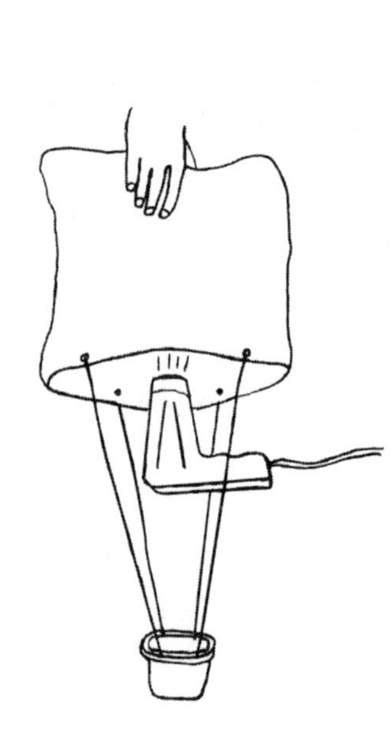

How to Proceed: Create a large-volume bag out of very thin plastic. Tie a small plastic container to the plastic bag with strings as shown. Inflate the bag with hot air from a hairdryer held just below the bag opening. Release the balloon.

- Did the balloon rise?
- How long did it stay in the air?
- Will the balloon lift weight placed in the plastic container?
- Can the balloon be made with tissue or other materials?

Activity 3. **Make Several Structures that Can Fly**

STUDENT EXPECTATION

Design, construct, and test a structure that can fly.

How to Proceed:

A. ***To Make a Parachute.*** Suspend metal washers from strings tied to the corners of a 250 mm square of thin plastic. Fold the parachute into a ball and throw it high into the air. Does it float gently to the ground? Does varying the size of the parachute or number of washers change the way it floats to the ground? Does cutting a hole in the top of the parachute canopy affect the way it floats to the ground?

B. ***To Make a Hang Glider.*** Make an Art Straw triangle secured on each side of the corners with gussets. Cut a triangle of thin plastic allowing an overlap to be folded over the straws and taped to itself. Loop a short piece of thread over the centre straw and attach a small ball of plasticine to the end. Temporarily tape the thread loop to the centre straw. Does the glider fly? Does it have a gently descending flight? What happens to its performance if the weight is shifted? Can it be propelled (Elastic band launcher, balloon propulsion, plastic squeeze bottle)?

C. ***To Make a Helioblade.*** Cut a rectangular strip form the side of a plastic bottle. Find the centre of the strip and drill or punch a hole slightly smaller than a short length of dowel rod. Force the dowel rod through the hole. The plastic strip must be a tight fit on the dowel. Twist the blades of the strip slightly. With the rotor blade facing upward, spin the dowel between the palm of your hands and throw the Helioblade in the air. What happens if you reverse the direction of blade spin? Are there other blade sizes or shapes that are more efficient? What would happen with two or more blades on the dowel?

Activity 4. **Exploring Ways to alter Drag**

STUDENT EXPECTATION

Demonstrate and describe methods used to alter drag in flying devices.

How to Proceed: Fold a sheet of paper as shown. Glue-stick the folds in place. Weight the nose with paper clips or Blue-tack. Alter the flaps at the end of the wing to see how it affects the flight path. What happens with both flaps up? What happens with both flaps down? What happens with one flap up and the other flap down?

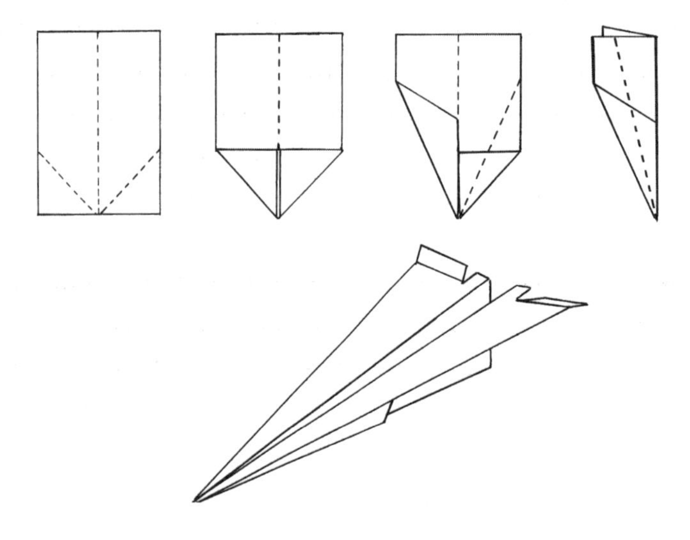

STRAND: Matter and Materials
TOPIC: Fluids
LEVEL: Grade 8

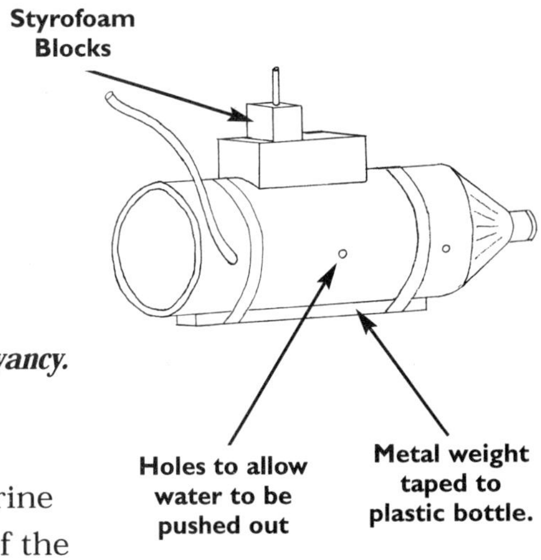

DEVICES TO SINK OR FLOAT

STUDENT EXPECTATION

Recognize and state the relationship between gravity and buoyancy.

Styrofoam Blocks

Holes to allow water to be pushed out

Metal weight taped to plastic bottle.

Activity 1 ***Make a Submarine.*** Fill the plastic bottle (submarine) with water. Place the submarine in a deep basin of water. Adjust the size of the weight until a portion of the tower is above the water. Use a large syringe as a pump to push the water from the submarine. This should raise the submarine in the water (air has replaced the water). Can the submarine be made to go completely below the surface of the water?

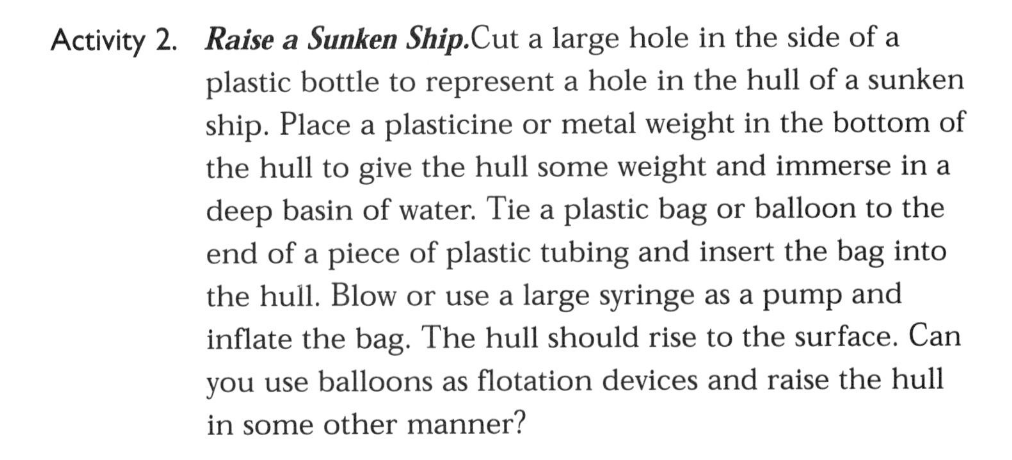

Activity 2. ***Raise a Sunken Ship.***Cut a large hole in the side of a plastic bottle to represent a hole in the hull of a sunken ship. Place a plasticine or metal weight in the bottom of the hull to give the hull some weight and immerse in a deep basin of water. Tie a plastic bag or balloon to the end of a piece of plastic tubing and insert the bag into the hull. Blow or use a large syringe as a pump and inflate the bag. The hull should rise to the surface. Can you use balloons as flotation devices and raise the hull in some other manner?

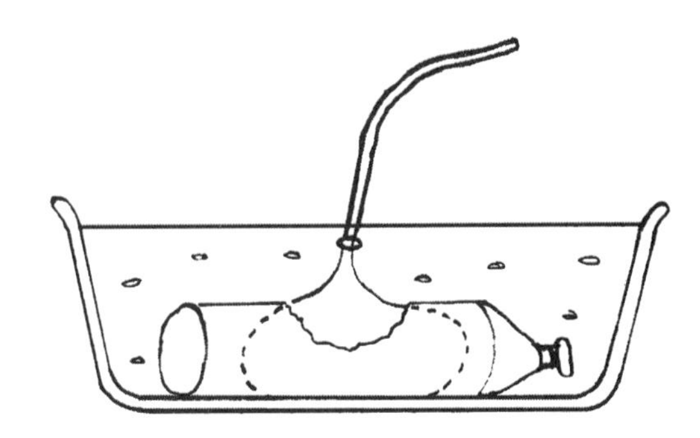

STRAND: Energy and Control
TOPIC: Energy From Wind and Moving Air
LEVEL: Grade 2

DEVICES PROPELLED BY AIR

STUDENT EXPECTATION

Design and construct a device propelled by air.

SAFETY FIRST!
Do not fly kites near overhead electrical wires.

Activity 1. ***Make a Kite.*** Light plastic sheet (garbage bags), dowel rods and string will provide the materials needed for a variety of kite designs. Try a variety of shapes (square, rectangle, triangle). What is the smallest kite that can be made to still fly? Try competitions to make a kite fly the highest. What is the heaviest load that can be lifted with a kite? Are there a variety of methods that can be used to fasten the string to the kite to get control? Is a tail necessary? What can be used to make a tail? Is the length of the tail important? Why does the kite rise into the air?

Activity 2. ***Make a Windmill.*** Cut a sheet of paper 150 mm square. From each corner, cut slits 55 mm. Fold the four corners as shown above, and put a straight pin through the folds. Tap the pin into a wood strip that can be used as a wand. A short length of Art Straw between the wood strip and windmill will keep the windmill from rubbing on the wand.

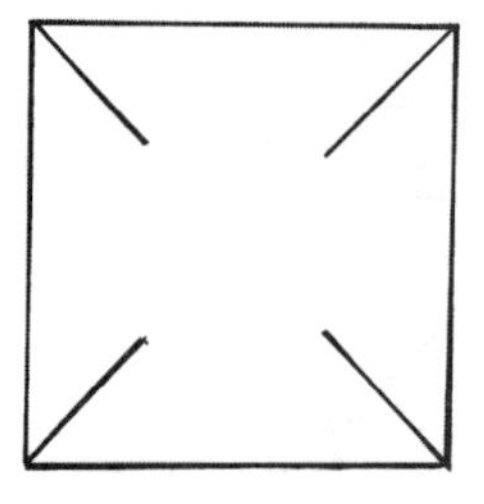
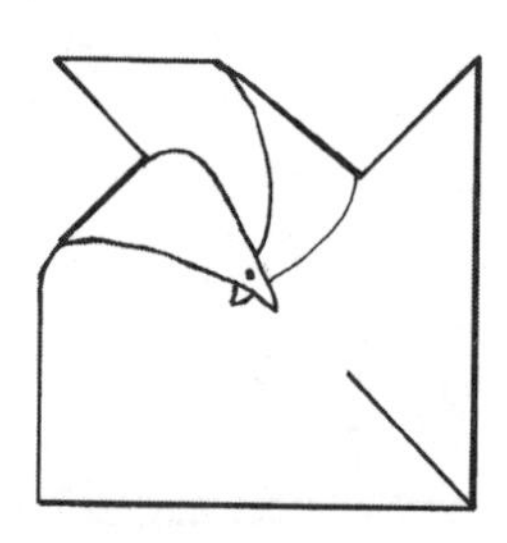
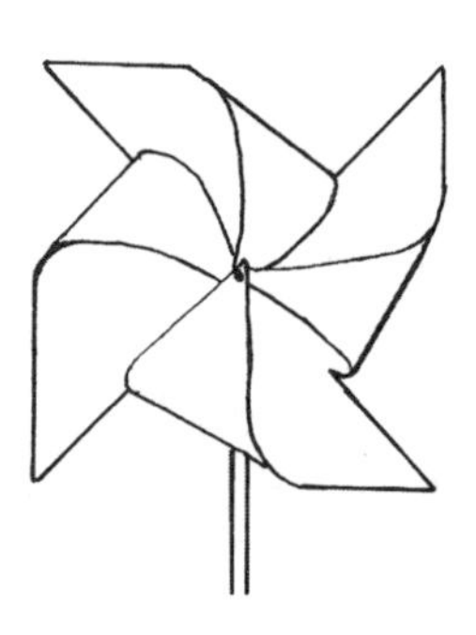

STRAND: Energy and Control
TOPIC: Forces and Movement
LEVEL: Grade 3

MAKE AN AIR-PROPELLED BOX

STUDENT EXPECTATIONS

Investigate the effects of directional forces and how unbalanced forces can cause visible motion in objects that are capable of movement.

Design and Construct a device that uses a specific form of energy in order to move.

Identify surfaces that affect the movement of objects by increasing or reducing friction.

Activity 1. **Make the Device:** Strap a blown-up balloon in a shoe box with the neck of the balloon projecting out through a small hole cut in the end of the box. Try propelling the box across a variety of floor surfaces (tile, carpet). Try propelling the box on top of Art Straws used as rollers.

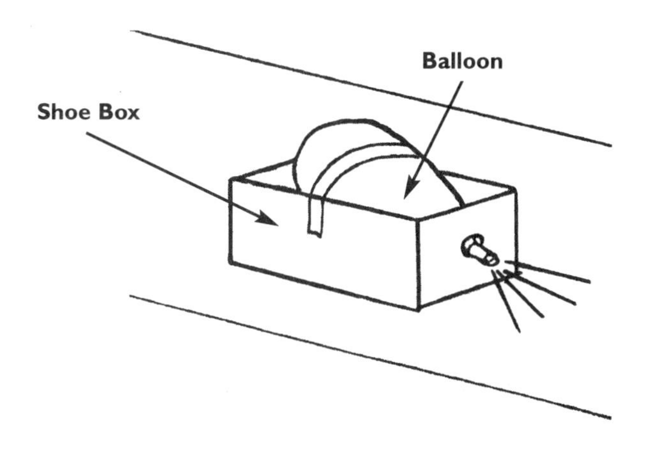

STRAND: Energy and Control
TOPIC: Light and Sound Energy
LEVEL: Grade 4

LIGHT AND SOUND

Activity 1. **Make Paper Shadow Images**

STUDENT EXPECTATION

Predict the location, shape, and size of a shadow when a light source is placed in a given location relative to an object.

How to Proceed: Cut various paper shapes. Darken the classroom as much as possible and using a lantern or flashlight, project the images onto a screen or wall. Have students look for a relationship between the size of the paper shape and the projected image size, while moving the shapes closer to, and further from, the light source. Try projecting the paper shapes to the screen using an overhead projector. How is the horizontal image projected to a vertical surface?

Activity 2. **Make a Periscope**

STUDENT EXPECTATIONS

Investigate and compare how light interacts with a variety of optical devices.

Design, make, and test an optical device.

How to Proceed: Tape a mirror on each end of a length of cardboard tube. Cut viewing windows, in the sides of the tube, close to the mirrors. What angle do the mirrors have to be placed on each end of the tube for proper viewing?

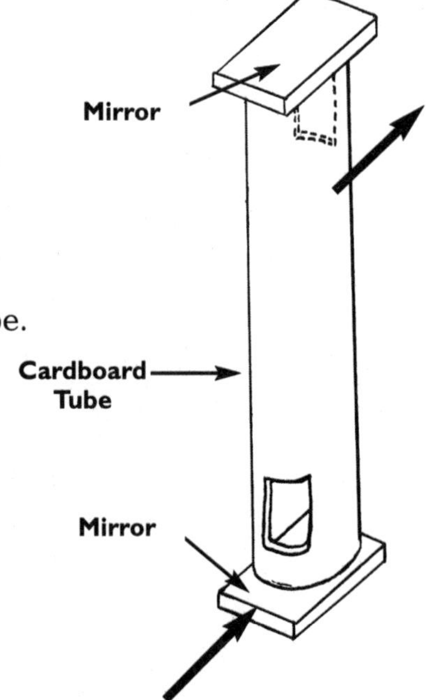

Activity 3. **Build Different Musical Instruments**

STUDENT EXPECTATIONS

Recognize that sounds are caused by vibrations.

Design and make musical instruments, and explain the relationship between the sounds they make and their shapes.

How to Proceed:

A. ***Stringed Instrument.*** Stretch Elastic Bands over a Margarine tub or cardboard box. With a hole cut in the top to make the tub or box into a sounding box.

B. ***Wind Instrument.*** Tape 3 or 4 different lengths of 12.5 mm inside diametre (½ inch) rigid PVC pipe together. Cork the bottom ends of each pipe. Blow across the top of the pipes to create sound (Figure 1).

Make a wind instrument to be able to adjust the sound it makes. Sand a cork to move freely within a piece of PVC pipe. Fasten the cork to the end of a piece of dowel. Blow across the open end of the pipe while moving the cork up and down the pipe (Figure 2).

C. ***Percussion Instrument.*** Try tapping cans, saucepans, cardboard boxes, foil-covered plastic tubs, or liquid-filled bottles with a stick. Glue wood strips spaced along the length of a cardboard box and drag a piece of dowel across the pieces.

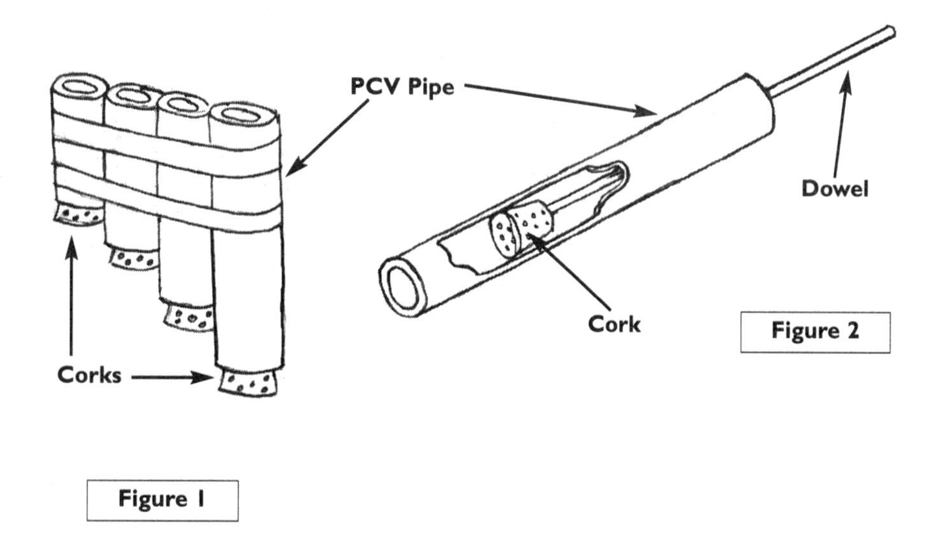

Figure 2

Figure 1

STRAND: Structures and Mechanisms
TOPIC: Everyday Structures
LEVEL: Grade 1

MAKING VARIOUS STRUCTURES WITH SPECIFIC FUNCTIONS

STUDENT EXPECTATION

Design and make different structures using concrete materials, and explain the function of the structure.

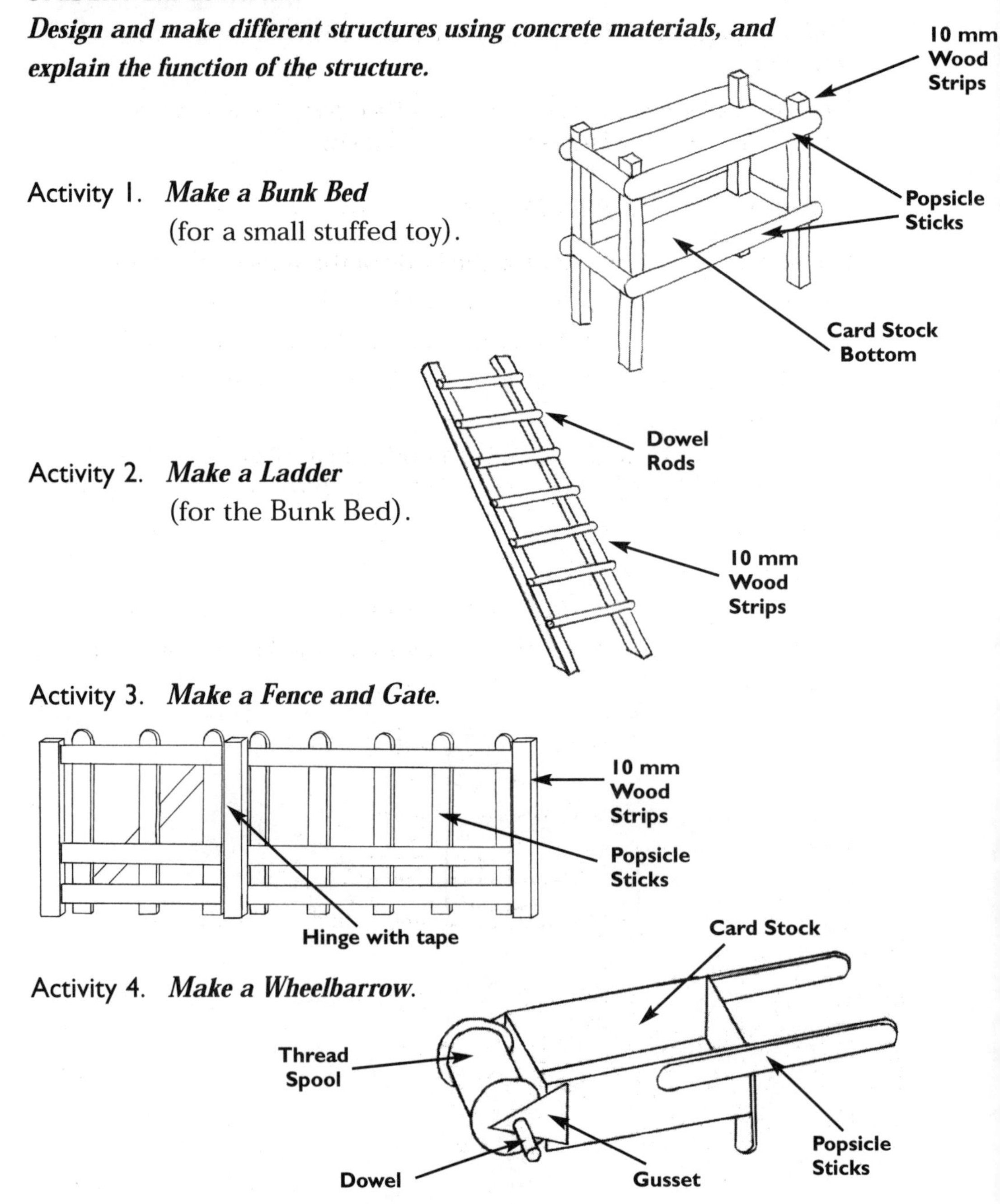

Activity 1. ***Make a Bunk Bed*** (for a small stuffed toy).

Activity 2. ***Make a Ladder*** (for the Bunk Bed).

Activity 3. ***Make a Fence and Gate.***

Activity 4. ***Make a Wheelbarrow.***

STRAND: Structures and Mechanisms
TOPIC: Movement
LEVEL: Grade 2

MAKING "SOMETHING" OUT OF "NOTHING"

STUDENT EXPECTATIONS

Describe, using their observations, the characteristics and movements of simple mechanisms.

Make simple mechanisms and use them in building a device they have designed.

Describe, using their observations, the effect that different surfaces have on the rate at which an object slows down.

Activity 1. **Make a Junk-Mobile**

Use a foam food container for the body of the vehicle. Demonstrate different ways to fasten axles and wheels on the vehicle. (Review "Wheels and Axles" in the Machines and Mechanisms section).

- What modifications can be made if the wheels are rubbing against the body of the vehicle?
- Does the vehicle run freely on both tile and carpet floors?
- Run the vehicle down a ramp. How does the ramp angle affect the distance the vehicle travels on the floor? (See Appendix D for recording distances)
- Add weight in the food container used as the vehicle body. Does the weight affect the distances traveled at the previously recorded angles?

SAFETY FIRST!

Because of the concerns regarding salmonella food poisoning, please thoroughly wash any containers that have held food (styrofoam trays, aluminum trays, cans) before sending them to class.

STRAND: Structures and Mechanisms
TOPIC: Stability
LEVEL: Grade 3

FACTORS AFFECTING THE STRENGTH AND STABILITY OF STRUCTURES

Activity 1. **Build a Beam Bridge**

STUDENT EXPECTATION

Describe, using their observations, ways in which the strength of different materials can be altered.

How to Proceed: Cut some cardstock into 600 mm x 150 mm strips. Space two thick books or house bricks approximately 350 mm apart and have students compare the strength of a flat strip, with a couple of coins set on it, to strips folded in a variety of ways (L-shape, U-shape, curve-shape, accordion-shape) while spanning the gap.

Activity 2. **Build a Tower**

STUDENT EXPECTATION

Describe ways to improve the strength and stability of a frame structure.

How to Proceed: Read the book *Architect of the Moon* to the class. In the book a message has arrived from the moon; "Help, I'm falling apart." Have students use Art Straws to build as tall a structure (tower) as possible to support the moon (a tennis ball). How can the structure be strengthened to keep it from twisting, bending, or collapsing? Why is vertical balance in the structure required? Why is a wider base required in the structure? How long will the structure support the moon (tennis ball)?

Activity 3. **Make a Lever Puppet**

STUDENT EXPECTATION

Describe, using their observations, how simple levers amplify or reduce movement.

How to Proceed: Review the section on Levers and have students use card stock to animate characters and objects using the simple lever. How should the lever be hinged to get minimum and maximum movement of the animated parts? Try animating two separate parts at the same time (arms and legs, head and tail, eyes and ears, etc.).

Activity 4. **Build a Crane**

STUDENT EXPECTATION

Design and make a stable structure that contains a mechanism and performs a function that meets a specific need.

How to Proceed: Have students build a simple crane that uses a winch and pulleys to lift a load. (See "pulleys" and "winches" in the Machines and Mechanisms section)

STRAND: Structures and Mechanisms
TOPIC: Pulleys and Gears
LEVEL: Grade 4

INVESTIGATING PULLEYS AND GEARS

Activity 1. **Make a Pulley Board.**

STUDENT EXPECTATIONS

Describe, using their observations, the function of pulley systems and gear systems.

Demonstrate an awareness of the concept of mechanical advantage by using a variety of pulleys and gears.

How to Proceed: Obtain a 300 mm square piece of 20 mm thick plywood. Rule the piece off with 6 equal spaces on both its length and width. Where the lines cross, drill 5 mm holes to suit the diameter of a dowel rod. This will give you several alternative locations to place pulleys. You can use plastic pulleys of varying diameters or use thread spools. Flat cans (salmon or tuna) also make good pulleys, but holes will have to be drilled in these for the dowel rods (see page 37 for this procedure). Be sure all pulleys turn freely on the dowels. Wind an elastic band around a cluster of pulleys.

- Have students notice the rotation of the driver to driven pulley(s);
- Twist the elastic band and have students notice the direction of driver and driven pulleys;
- Can you make one cluster of pulleys drive another cluster?

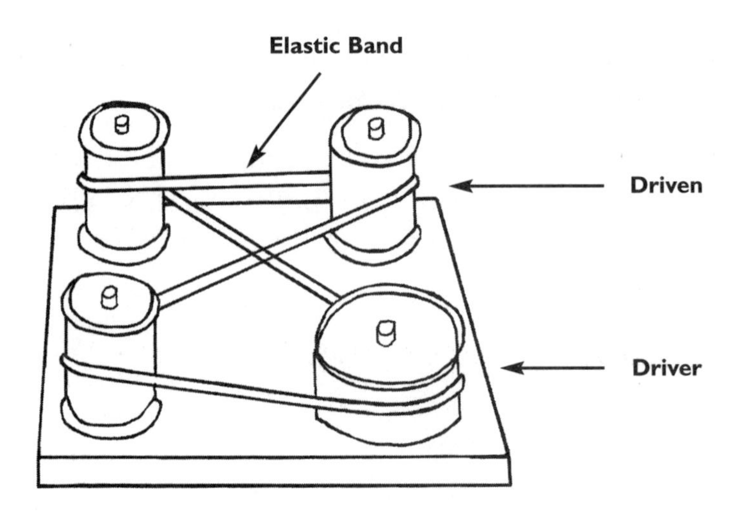

Activity 2. **Make a Gear Board**

STUDENT EXPECTATIONS

Describe, using their observations, the function of pulley systems and gear systems.

Demonstrate an awareness of the concept of mechanical advantage by using a variety of pulleys and gears.

How to Proceed: A Gear Board can be made much like the Pulley Board with the exception that holes for the dowel rods have to be carefully positioned so that each gear meshes with its adjoining gear. Use a variety of sizes of spur gears and place them in positions that will allow them to mesh properly and pencil the hole locations of each for drilling dowel positions.

- Have students count the rotations of adjoining gears when the driver gear is rotated 360 degrees.
- Have students think about a bicycle. Does driving the smaller or larger gear give the greater mechanical advantage?

Activity 3. **Make a Crane**

STUDENT EXPECTATION

Design and make a pulley system that performs a specific task.

How to Proceed: Make a crane using wood strips. Use a thread spool for the winch, plastic pulleys and a string cable.

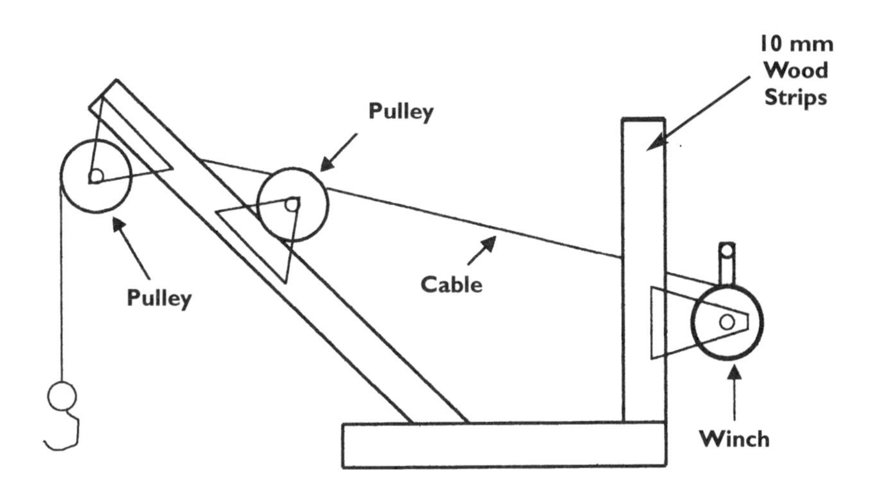

STRAND: Earth and Space Systems
TOPIC: Weather
LEVEL: Grade 5

DESIGNING AND BUILDING WEATHER INSTRUMENTS

STUDENT EXPECTATION

Design, construct and test a variety of weather instruments.

Activity 1. ***Build a Weather Vane.*** Students can use their own creativity in building a weather vane. The vanes can be tested in front of a fan. The weather vane shown has a weighted margarine tub with an eraser pencil inserted in it. A pin is pushed through an Art Straw and inserted in the eraser. The vanes are cut from card stock and inserted into slits cut in the straws.

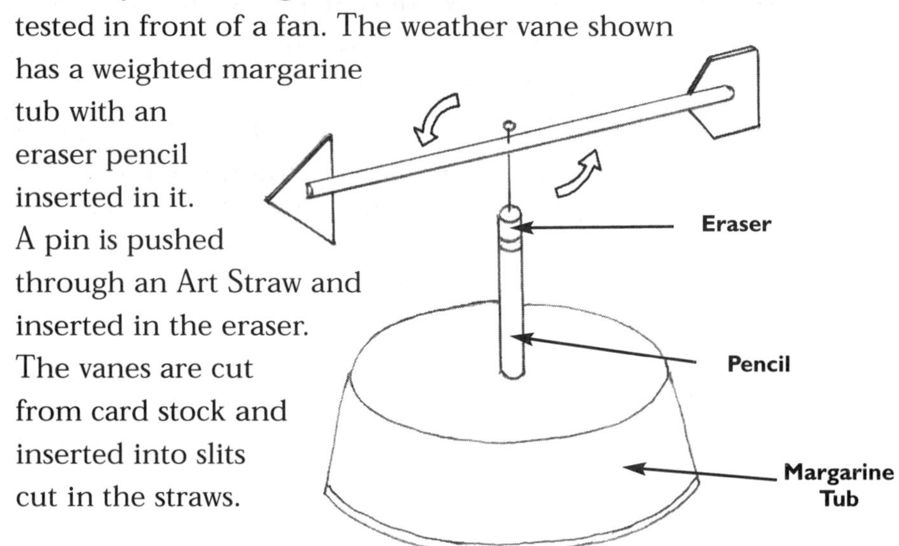

Activity 2. ***Build an Anemometer.***

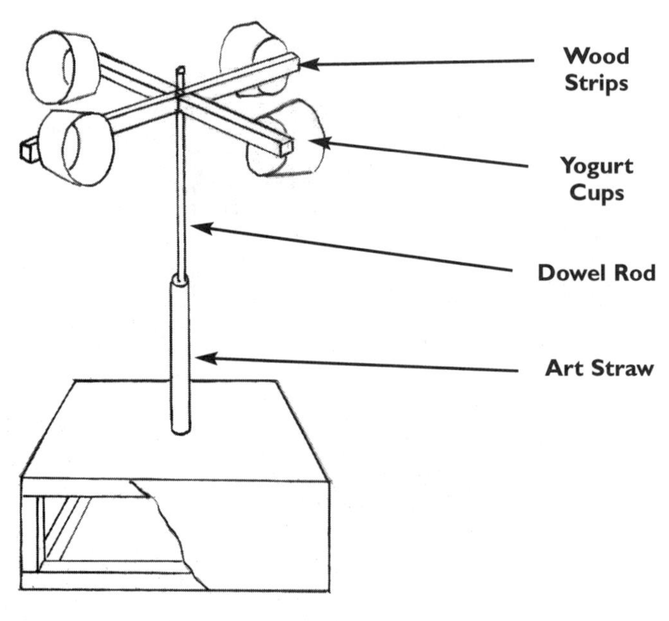

STRAND: Earth and Space Systems
TOPIC: Space
LEVEL: Grade 6

CONSTRUCTING A DEVICE TO TELL TIME

STUDENT EXPECTATION

Construct a device that could have been used to tell time before mechanical clocks were invented.

Activity 1. ***Build a Water Clock.*** Make a small hole with a pin in the bottom of each of the small plastic cups. Thumbtack each of the small cups, one below the other, to the stand. Place a larger container on the base of the stand to catch the water as it drains from the smaller cups. Fill the top cup with water. Measure the time it takes for the water to empty from the last of the small cups.

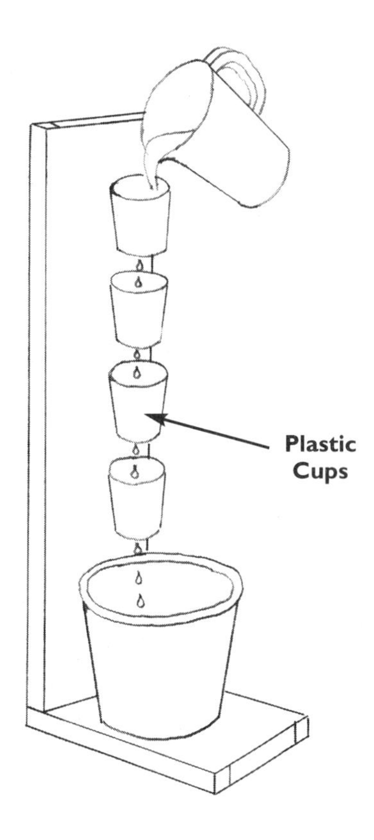

OPEN-ENDED PROBLEM-SOLVING CHALLENGES

The following activities are still tied to the specific expectations for students, but rather than being teacher directed, are open-ended problems for students to solve. Each is developed through a "situation" that sets the stage for the challenge. Each challenge is introduced with student expectations preceding the challenge, but it is by no means a complete list. Teachers are encouraged to develop more challenges that that are linked to the grade, curriculum strand, or student expectation not described in the following sample challenges.

Teachers will need to decide if, as presented, the challenges are age or ability appropriate. This does not mean that some younger students would not be able to design and build the challenge, but they may not yet be able to reason through the formal design process. For those students, you will be able to lift out the challenge itself and present it as a teacher directed specific activity as before.

For students who are able to apply the open-ended problem-solving process to these challenges, the Design Process is encouraged. Within each challenge, students are encouraged to consider:

NEED	IDEAS	CREATE	EVALUATE
What do we have to do?	How will we do it?	How can we make it?	How well did we do?

Note: Refer to the N. I. C. E. model on page 12 for a more complete description of the problem-solving process.

Following each challenge, a set of "Teacher Notes" will assist with a focus and some ideas to consider. These ideas should be for the teacher's eyes only and used to assist students with some direction if the problem-solving process has bogged down.

A SUMMARY OF THE CHALLENGES

Correlated by Strand, Topic and Grade to a sample curriculum

Strand	Topic	Grade Level	Challenge
Matter & Materials	Materials that transmit, reflect, or absorb light or sound	4	Design and build a simple communication system to be able to talk to your friend.
Matter & Materials	Materials that transmit, reflect, or absorb light or sound	4	Design and build a 3-string instrument that can be tuned.
Matter & Materials	Properties of and Changes in Matter	5	Design and build a container to slow the melting of an ice cube (from making a complete change of state.)
Matter & Materials	Properties of air and characteristics of flight	6	Design and build a self-propelled device to perform a set task.
Matter & Materials	Fluids	8	Design and build a model pneumatic or hydraulic device to perform a set task.
Energy & Control	Conservation of Energy	5	Design and build a timing device.
Energy & Control	Electricity	6	Design and build different types of electrical switches to perform set tasks.
Energy & Control	Electricity	6	Design and build both a series and a parallel circuit for a battery-operated string of lights.
Energy & Control	Heat	7	Design and build a container to minimize the melting of ice cubes surrounding 24 cans of soft drinks.
Structures & Mechanisms	Forces acting on structures	5	Design and build a bridge to satisfy contest parameters.
Structures & Mechanisms	Motion	6	Design and build a vehicle to perform set tasks.
Structures & Mechanisms	Structural strength and stability	7	To design and build a model crane to investigate why cranes don't topple under a load.
Structures & Mechanisms	Mechanical efficiency	8	Build a model of a piece of construction machinery to demonstrate "mechanical efficiency"

STRAND: Matter and Materials
TOPIC: Materials That Transmit, Reflect, or Absorb Light or Sound
LEVEL: Grade 4

COMMUNICATING

STUDENT EXPECTATION

Identify a variety of materials through which sound can travel.

Situation

Your teacher has reminded you that shouting to your friend across the room is not to continue. Even though the teacher has suggested that this shouting is disturbing to others, you would like to continue talking to your friend.

Challenge

Design and build a simple communication system to be able to talk to your friend.

Tin or Plastic Containers

Fishing Line, Wire, Elastic Band

Consider the following

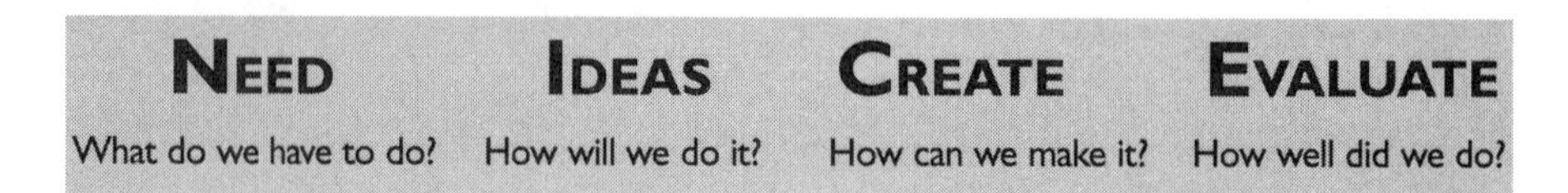

Tools and Materials

- Soup Cans
- Plastic Yogurt Containers
- Paper Cups
- Cotton String
- Nylon Fishing Line
- Thin Music Wire
- Elastic Bands
- Pliers
- Drill/Drill Bits

Teacher Notes

A length of string, fishing line, music wire, or elastic band stretched between soup cans, plastic containers, or paper cups is an obvious solution to this challenge.

- Is sound transmitted better through one material than another?
- Is sound sent and received better using paper, plastic, or metal containers?
- Is there an optimal length of wire etc. for voice travel to be most effective?

STRAND: Matter and Materials
TOPIC: Materials that Transmit, Reflect, or Absorb Light or Sound
LEVEL: Grade 4

AN INSTRUMENT THAT CAN BE TUNED

STUDENT EXPECTATION

Investigate and describe physical changes in a material that can alter the sound it makes

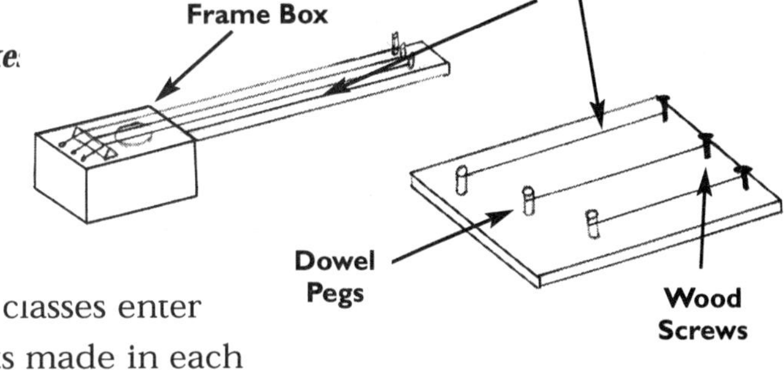

Situation

Your school has decided to put on a music concert. One of the challenges teachers have been given is to have their classes enter a talent contest using musical instruments made in each classroom. Your class has decided to enter the contest.

Challenge

Design and build a three-string instrument that can be tuned.

Consider the following

NEED	IDEAS	CREATE	EVALUATE
What do we have to do?	How will we do it?	How can we make it?	How well did we do?

Tools and Materials

25, 50, 900 mm Board
38 mm (closest to 1½ inch) Finishing Nails
Drills and Drill Bits
Thin Music Wire
Masking Tape
Hammer
Dowel Rod
Glue
Tape Measure/Ruler
Wood Screws
Pliers
Fishing Line
Saw/Bench Hook

Teacher Notes

The size or design of the instrument is not as important as the means to adjust the tension on the strings.

- Can the dowel pegs be adjusted for tuning?
- Can wood screws be used for the adjustment?
- Is fishing line as effective as music wire for the strings?

STRAND: Matter and Materials
TOPIC: Properties of and Changes in Matter
LEVEL: Grade 5

KEEPING IT COOL

STUDENT EXPECTATION

Recognize, on the basis of their observations, that melting and evaporation require heat.

Situation

Several teams of students within your classroom have been challenged to keep a single cube of ice from melting completely for as long as possible and to chart the melting times using a variety of insulation materials.

Challenge

Design and build a container to slow the melting of an ice cube (from making a complete change of state).

Consider the following

NEED	IDEAS	CREATE	EVALUATE
What do we have to do?	How will we do it?	How can we make it?	How well did we do?

Tools and Materials

Wood Strips	Glue	Tape	Scissors
Mitre Box/Saw	Sawdust	Paper	Ruler
Styrofoam Chips	Peat Moss	Ice Cubes	
Card Stock	Vermiculite	Plastic Containers	

Teacher Notes

The insulation quality of a variety of materials is important to this challenge.

- Should the insulating material surround (touch) the ice cube?
- Should the ice cube be placed in a separate container within another insulated container?
- Can a heat source be used to provide a fair test for each solution to the challenge and to speed up the melting process?
- How can the start and melted points of the ice cube be determined and charted fairly between solutions?
- What change of state is taking place?

STRAND: Matter and Materials
TOPIC: Properties of Air and Characteristics of Flight
LEVEL: Grade 6

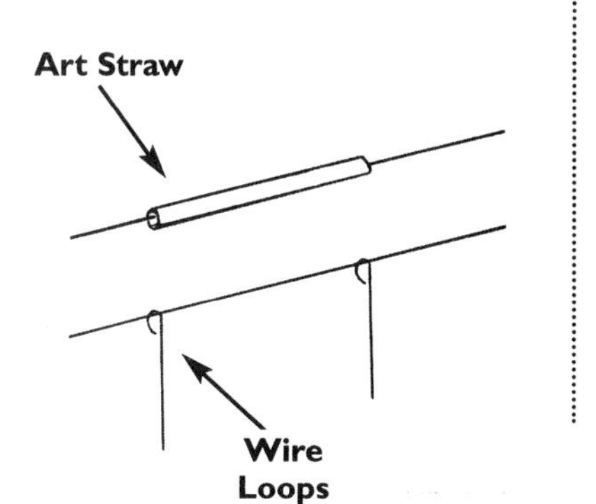

FLYING ACROSS THE GYM

STUDENT EXPECTATION

Describe the sources of propulsion for flying devices.

Situation

Several classes within your school have decided to accept a challenge given by the Physical Education teacher, to propel a small cloth replica of the school mascot (animal, bird, etc.) as far as possible along a wire stretched from one side of the gymnasium to the other. The mascot will signal the introduction of the team players and can sit, hang, or lie down on a platform suspended from the wire.

Challenge

Design and build a self-propelled device that will transport a platform holding the small cloth mascot as far along a suspended wire as possible. The platform is to be no longer than 200 mm and no wider than 100 mm.

Consider the following

NEED	IDEAS	CREATE	EVALUATE
What do we have to do?	How will we do it?	How can we make it?	How well did we do?

Tools and Materials

Mascot (cloth animal)	Balloons	Glue
Wood strips	Propellers	Art Straws
Card stock	Elastic bands	Needle Nose Pliers
Wire	Tape	Ruler

Teacher Notes

- The wire can be replaced with nylon fishing line for a classroom application.
- Platform should be as light as possible (wood strips covered with card stock?)
- How will the platform be run along the wire? (balloon-propulsion, elastic-driven propeller, motor-driven propeller, etc.)
- How will the platform be suspended from the wire or nylon fishing line? (Art Straw, wire loops, etc.)

STRAND: Matter and Materials
TOPIC: Fluids
LEVEL: Grade 8

GETTING A LIFT

STUDENT EXPECTATIONS

Compare fluids in terms of their compressibility or incompressibility.

Compare liquids and air in terms of their efficiency as transmitters of force in pneumatic and hydraulic devices.

Situation

A small engineering company close to your school has asked your class to come up with some ideas on how to raise a platform of containers of scrap materials from the floor to a raised deck level.

Challenge

Design and build a model pneumatic or hydraulic device to raise a 100 mm square platform with 4 film canisters filled with sand (containers of scrap materials), from the floor to an upper deck level 150 mm high.

Consider the following

NEED	IDEAS	CREATE	EVALUATE
What do we have to do?	How will we do it?	How can we make it?	How well did we do?

Tools and Materials

Wood Strips	Saw	Scissors
Popsicle Sticks	Glue	Drill/Drill Bits
Dowel Rods	Tape	Syringes/Tubing
Card Stock	Bench Hook	Film Canisters (filled with sand)

Teacher Notes

- Review the section on Energy and Control.
- Lifting the platform with a device on the floor level is one alternative.
- Lifting with the device placed on the deck level is another alternative.
- How should the platform be made so that a lifting device can get under it to lift?
- Is the system more efficient with pneumatics or hydraulics (system with or without water in the syringes and tubing)? Why?

STRAND: Energy and Control
TOPIC: Conservation of Energy
LEVEL: Grade 5

MEASURING MINUTES

Buzzer
Battery Pack
Thumb Tacks
Cork
Plastic Containers

STUDENT EXPECTATION

Operate a mechanical device or system that uses sensory or time-based input.

Situation

You are on a camping trip with your class. You are without clocks or watches and have been challenged to make a device that will signal when the eggs, being boiled for breakfast, are cooked.

Challenge

Design and build a timing device that will signal when the eggs have been boiling for two minutes.

Consider the following

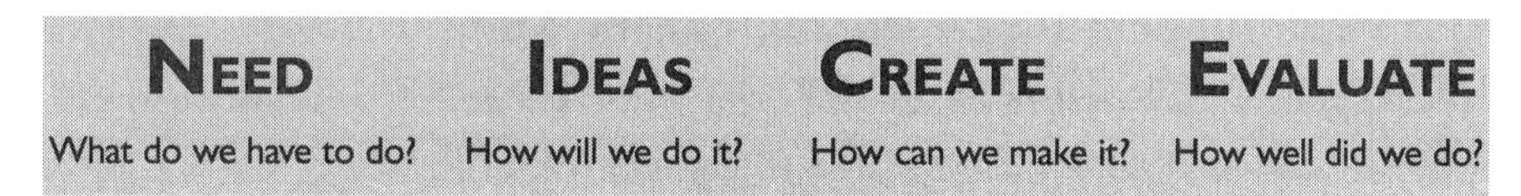

Tools and Materials

Wood Strips	Scissors	Thumb Tacks
Popsicle Sticks	Glue	Drill/Drill Bits
Dowel Rods	Tape	Saw
Bench Hook	Card Stock	Water
Bulb/Bulb Holder	Buzzer	Wire
Batteries/Battery Holder	Corks	Plastic Containers

Teacher Notes

This challenge has many solutions (sand-timer, calibrated burning candle, water-clock). Raising a flotation device (cork) or lowering the float to make electrical connection to a signaling device in some way is the challenge.

STRAND: Energy and Control
TOPIC: Electricity
LEVEL: Grade 6

SWITCHING AROUND

STUDENT EXPECTATION

Identify different types of switches that are used to control electrical devices.

Situation

You have built an electrical circuit but find that all the available switches are in use.

Challenge

Design and build at least three different types of electrical switches that will allow you to:

A. Turn a light on and off;
B. Simulate ringing a door buzzer.

Consider the following

NEED	IDEAS	CREATE	EVALUATE
What do we have to do?	How will we do it?	How can we make it?	How well did we do?

Tools and Materials

Wood Strips	Buzzer	Art Straws
Card Stock	Popsicle Sticks	Tin Foil
Dowel Rods	Batteries	Wire
Thumb Tacks	Glue	Scissors
Tape	Nails/Pins	Paper Clips
Ruler	Saw	Bench Hook
Bulb/Bulb Holder		

Teacher Notes

The paper clip switch is a good starting point for this challenge. Students need to be shown that a paper clip is a good electrical conductor. To show this, fasten a paper clip in a circuit between a battery and light bulb. Try other materials as electrical conductors (coins, foil, coat hangers, bar magnets, etc.)

STRAND: Energy and Control
TOPIC: Electricity
LEVEL: Grade 6

LIGHT CIRCUITS

STUDENT EXPECTATIONS

Design and build a series circuit and a parallel circuit and describe the function of their component parts.

Build and test an electrical circuit that performs a useful function, and draw a diagram of it using appropriate electrical symbols.

Situation

Your class has been asked to provide a battery-operated string of lights in a display case in the school lobby.

Challenge

Design and build both a series and parallel circuit and draw a circuit diagram for each.

Consider the following

NEED	IDEAS	CREATE	EVALUATE
What do we have to do?	How will we do it?	How can we make it?	How well did we do?

Tools and Materials

Wood Strips	Wire	Wire Stripper
Bench Hook	Batteries/Battery Holder	Ruler
Card Stock	Bulbs/Bulb Holders	Scissors
Paper Clips	Thumb Tacks	Saw

Teacher Notes

Review "Electrical Energy" in the Energy and Control section. Students need practice with circuit diagrams and the use of electrical symbols. As an initial attempt at understanding circuit diagrams, photocopy a set of symbols and have students choose and place these in a circuit.

STRAND: Energy and Control
TOPIC: Heat
LEVEL: Grade 7

KEEPING THE ICE

STUDENT EXPECTATION

Design and build a device that minimizes energy transfer.

Situation

Once again, your class has been asked to provide soft drinks for the Track and Field event the school is hosting. Your teacher indicated that last year, it became difficult to keep the ice from melting in the open metal tubs that were used.

Challenge

Design and build a chest or box with a protective lid and an insulated lining that will minimize the melting of ice cubes surrounding 24 cans of soft drinks.

Consider the following

NEED	IDEAS	CREATE	EVALUATE
What do we have to do?	How will we do it?	How can we make it?	How well did we do?

Tools and Materials

Wood Strips	Foam Chips	Peat Moss
Cardboard	Tape	Glue
Ice Cubes	Masonite (Hardboard)	Vermiculite
Bench Hook	Foam Insulation Board	Ruler
Stapler/Staples	Polyethylene Sheet	Saw/Sawdust

Teacher Notes

This challenge may require students to cut some sheet materials (hardboard) or once the design has been determined, can be cut in local secondary schools. If students cut the material themselves, a fine-toothed crosscut saw will be needed. Lines that outline the pieces to be cut should be accurately marked out with a pencil and straight edge. Make sure students saw on the outside of the line. If foam board is used, it can be cut with a bread knife or with the crosscut saw. Hinges may be purchased or an alternative might be to use leather straps or dowel rods inserted through small screw eyes.

STRAND: Structures and Mechanisms
TOPIC: Forces Acting on Structures
LEVEL: Grade 5

CAN IT STAND?

STUDENT EXPECTATIONS

Design and make a frame structure that can support a load.

Identify the parts of a structure that are under tension and those that are under compression when subjected to a load.

Situation

The Soupson Soup Company is sponsoring a bridge-building contest that your class has decided to enter. The contest has a variety of categories that include bridges made of Pasta, Popsicle Sticks, or 10 mm Wood Strips. Each bridge is required to span a 450 mm opening between two tables and will be assessed on the number of cans of soup, suspended below the bridge, it can support until the bridge fractures.

Challenge

Choose one of the categories listed above and design and build a bridge.

Consider the following

NEED	IDEAS	CREATE	EVALUATE
What do we have to do?	How will we do it?	How can we make it?	How well did we do?

Tools and Materials

Wood Strips	Bench Hook	Glue Gun
Ruler	Variety of Pasta	Saw
Carpenter's Glue	Scissors	Popsicle Sticks
Cardstock	Tape	

Teacher Notes

- Depending on the materials available, teachers may decide to direct the category of bridge to be made and whether or not the bridge will be put to a destruct test.
- Which bridge members are under tension and which are under compression?

STRAND: Structures and Mechanisms
TOPIC: Motion
LEVEL: Grade 6

SLIDING TO SAFETY

STUDENT EXPECTATIONS

Investigate ways of reducing friction so that an object can be moved more easily.

Describe, using their observations, ways in which mechanical devices and systems produce a linear output from a rotary input.

Situation You have been shrunk to the size of a mouse (as in the movie *Honey, I Shrunk the Kids*). One day, three of you were mistakenly swept up and discarded into the trashcan outside the house. It's a long way down to the ground (too far to jump) and the house seems a long way back.

Challenge 1 • Using the materials found in the trashcan, work as a team to design and build a vehicle to carry you down a 150 mm wide wood plank leaning against the trashcan, and as close to the house as possible. Your vehicle must be less than 150 mm wide.

Consider the following

NEED	IDEAS	CREATE	EVALUATE
What do we have to do?	How will we do it?	How can we make it?	How well did we do?

Tools and Materials

Saw	Carpenter's glue	10 mm Wood Strips
Bench Hook	5 mm Dowel	Card Stock
Ruler	Art Straws	Tape

Challenge 2 • Now that you are on the ground, modify your vehicle to control its movement for the remaining distance to the house.

Additional Materials

Electric Motor	Motor Mount	Thumb Tacks
Battery Holder	Pulleys	Elastic Bands
Electric Wire	Paper Clips	

Teacher Notes A Vehicle Diagram and Record Chart is provided in Appendix C and D. Teachers will need to decide whether or not to share this diagram with students. Their problem-solving ideas should be considered.

STRAND: Structures and Mechanisms
TOPIC: Structural Strength and Stability
LEVEL: Grade 7

THE CRANE'S SECRET

STUDENT EXPECTATION

Demonstrate awareness that the position of the centre of gravity of a structure determines whether or not the structure is stable or unstable.

Situation

With all the high-rise building going on, your class was discussing the use of Building Cranes that are used to assist in the construction of the building. A question was asked by one of your classmates "with the crane boom sticking out so far on one side, why doesn't the crane doesn't topple over when it picks up a load?"

Challenge

To investigate the answer to the question and to investigate balance in a structure, design and build a model of a Building Crane that will lift a load and deposit it at 90 degrees to its original location.

Consider the following

NEED	IDEAS	CREATE	EVALUATE
What do we have to do?	How will we do it?	How can we make it?	How well did we do?

Tools and Materials

Wood Strips	Pulleys	Wood Wheels
Card Stock	String	Dowel Rod
Bench Hook	Carpenter's Glue	Paper Clips
Tape	Syringes/Tubing	Ruler
Saw	Balance Weights	

Teacher Notes

- A Typical Building Crane structure is included in Appendix F for teacher assistance.
- How can a centre of gravity of the structure be maintained?
- What gives a crane tower its stability?

STRAND: Structures and Mechanisms
TOPIC: Mechanical Efficiency
LEVEL: Grade 8

IMPROVING MACHINES

STUDENT EXPECTATIONS

Explain, using their observations, how the use of appropriate levers and ways of linking the components of fluid systems can improve the performance of the systems.

Predict the mechanical efficiency of using different mechanical systems.

Situation

The area in which your school is located has several major construction projects underway. Heavy equipment is being used, including: earth moving equipment, dump trucks, and building cranes.

Challenge

Build a model of one of the pieces of construction machinery to demonstrate "Mechanical Efficiency."

Consider the following

NEED	IDEAS	CREATE	EVALUATE
What do we have to do?	How will we do it?	How can we make it?	How well did we do?

Tools and Materials

Wood strips	Scissors	Drill/Drill Bits
Popsicle Sticks	Glue	Pulleys
Dowel Rods	Tape	Saw
Ruler	Card Stock	Bench Hook
String	Syringes/Tubing	

Teacher Notes

- Students may need to experiment with syringes of various sizes and linked with tubing in combination with one another.
- Which is the Master Cylinder and which is the Slave Cylinder?
- With different size syringes linked together, which cylinder is used as the Master Cylinder to give maximum plunger movement?
- What is the advantage of hydraulic systems over pneumatic systems?
- With a syringe used to lift a lever, is the syringe closer or further from the fulcrum to give minimum lever movement? Which requires more effort; placing the syringe closer or further from the fulcrum?

Section 12

Appendices

Appendix A

Please Save For Me

Dear Parent/Guardian:

Our children are asking for your help in saving some of the following materials for some exciting Science and Technology we will be doing this year. Thank you for your help.

Styrofoam	**Cardboard Tubes**	**Lids**
meat trays	waxed paper	jam jars
cups	foil	pickle jars
	paper towel	juice bottles
	toilet paper	
	wrapping paper	
Aluminum	**Plastic Bottles**	**Hardware**
pie tins	soda pop	nails
TV dinner trays	dish detergent	bolts
foil	water	nuts
		washers
Tin cans	**Cardboard Boxes**	**Spools**
soup salmon	cereal	ribbon
pop tuna	Kleenex	thread
juice		fishing line
Broken Appliances	**Film Canisters**	**Egg Crates**
toasters		
radios	**Clothes Pins**	
tools		

SAFETY FIRST!

Because of the concerns regarding salmonella food poisoning, please thoroughly wash any containers that have held food (styrofoam trays, aluminum trays, cans) before sending them to class.

BOX TEMPLATE

Appendix B

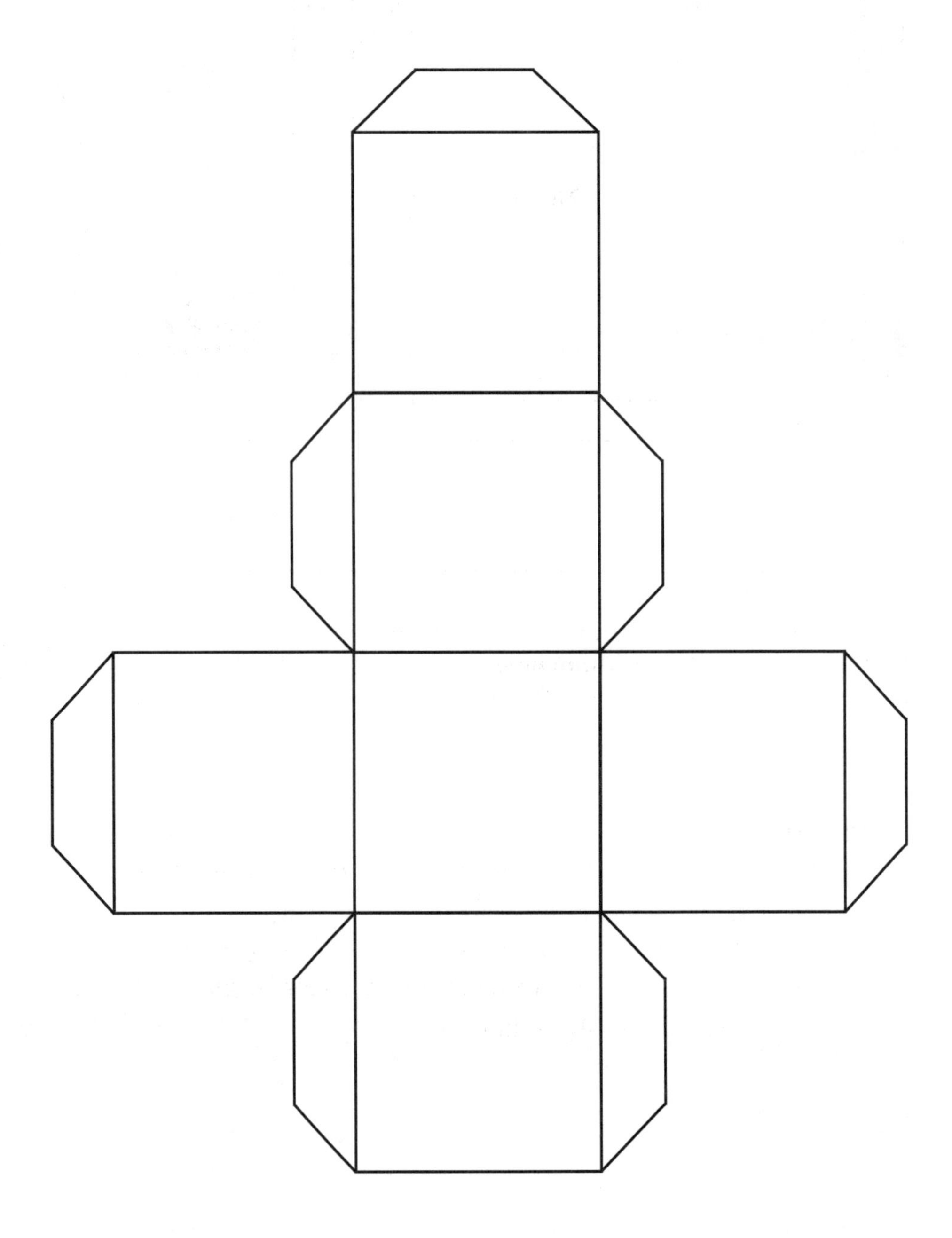

VEHICLE CHASSIS

Appendix C

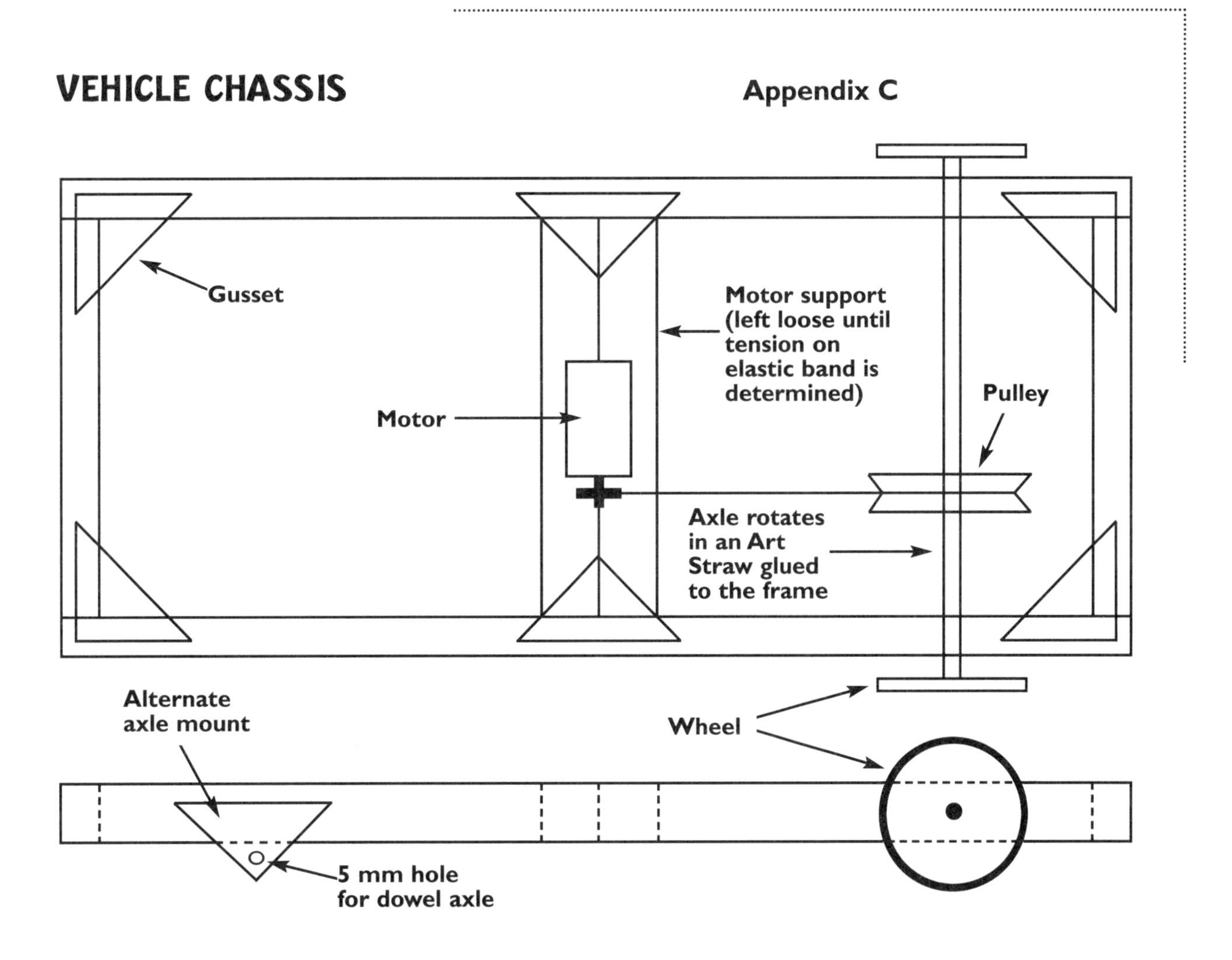

Alternate Axle Mounts

1. Mount "jumbo" Art Straw onto chassis for 5 mm dowel axle.

OR

2. Mount cardstock gusset on chassis. Punch hole for 5 mm dowel axle. (Gusset can be made stronger by gluing 3 or 4 gussets into multiple layers)

RECORD CHART

Appendix D

GRAVITY/FRICTION RELATIONSHIPS

The following will show the relationship of the pull of gravity with friction as a factor. Run vehicles down a ramp set at a variety of angles. Record the distances travelled beyond the ramp. Add weight to the vehicle and record the distances travelled with the ramp set at each of the previous angles.

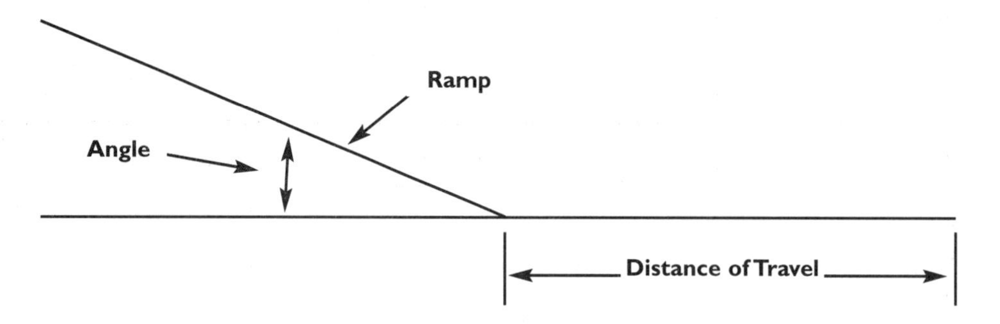

	Ramp Angle	Distance Travelled	Distance Travelled with Weight	Travelled Further? Y/N
1.				
2.				
3.				
4.				

Observations:

- Did the vehicle without weight travel further with greater ramp angles? Why?
- Did the weighted vehicle travel further than the weightless vehicle? If so, why? If not, why?
- Is there a relationship to the distance travelled, with and without the weight on the vehicle, to the higher and lower ramp angles? Explain what might be happening.

GUSSETS

Appendix E

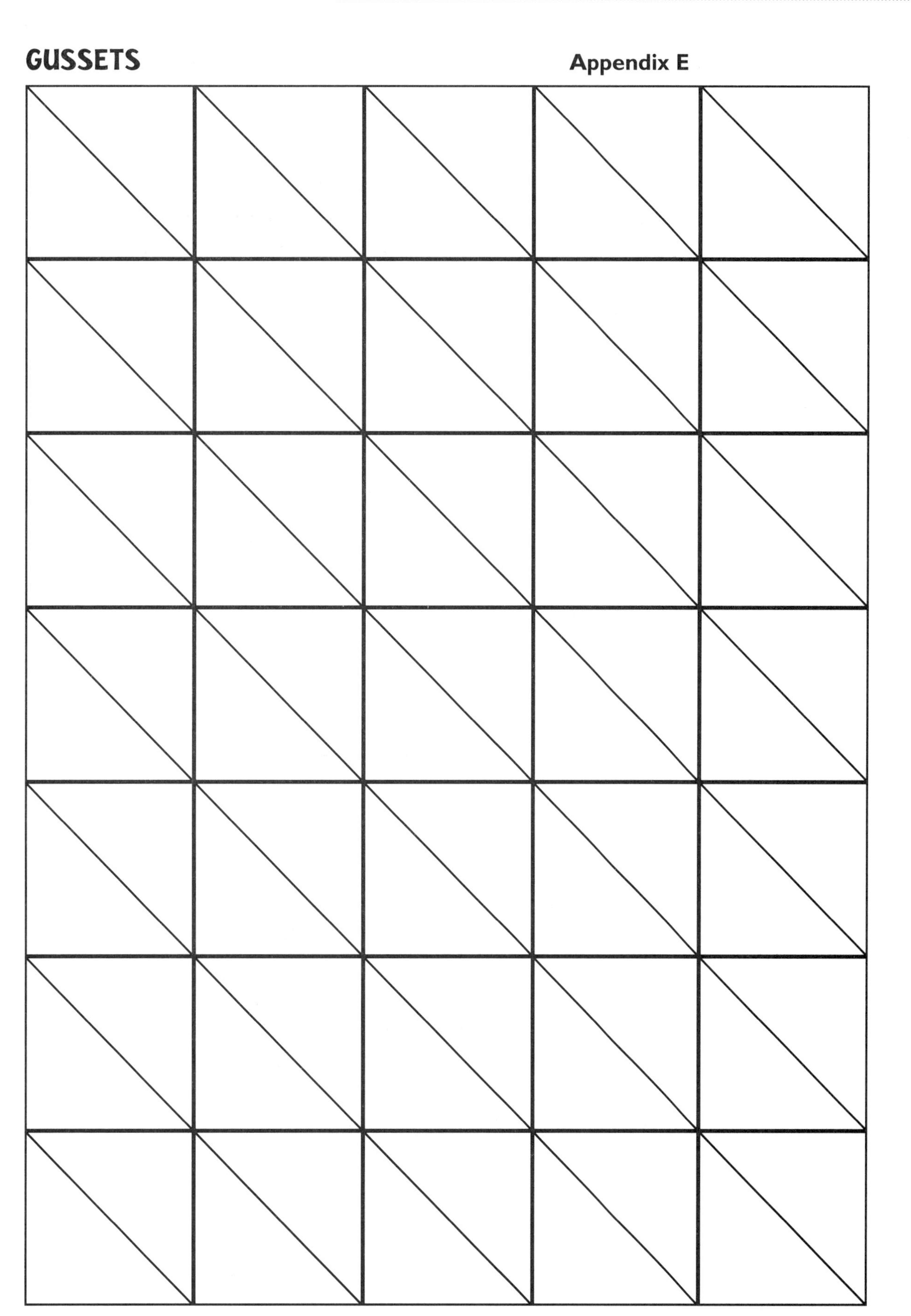

BUILDING CRANE

Appendix F

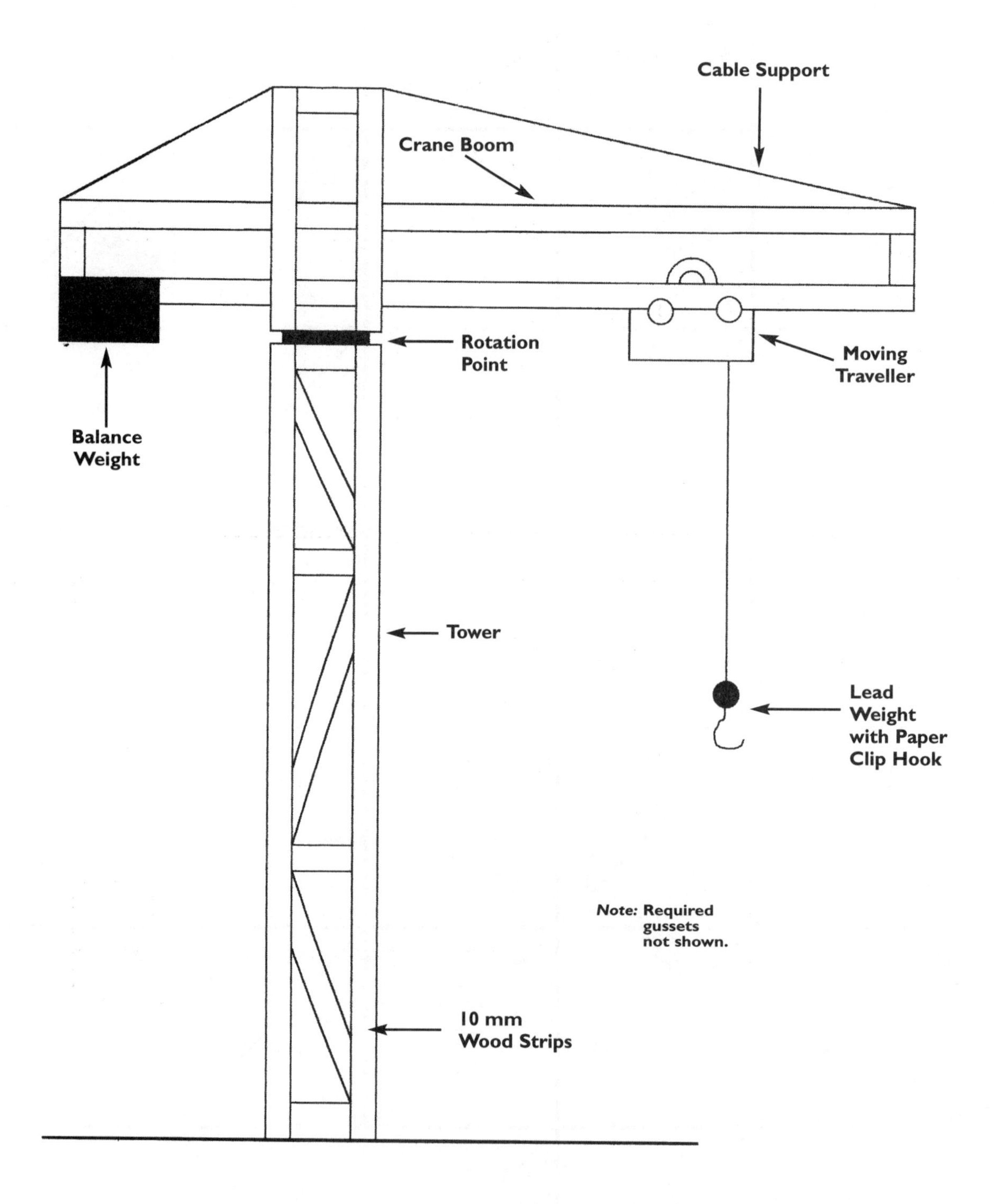

BENCH HOOK

Appendix G

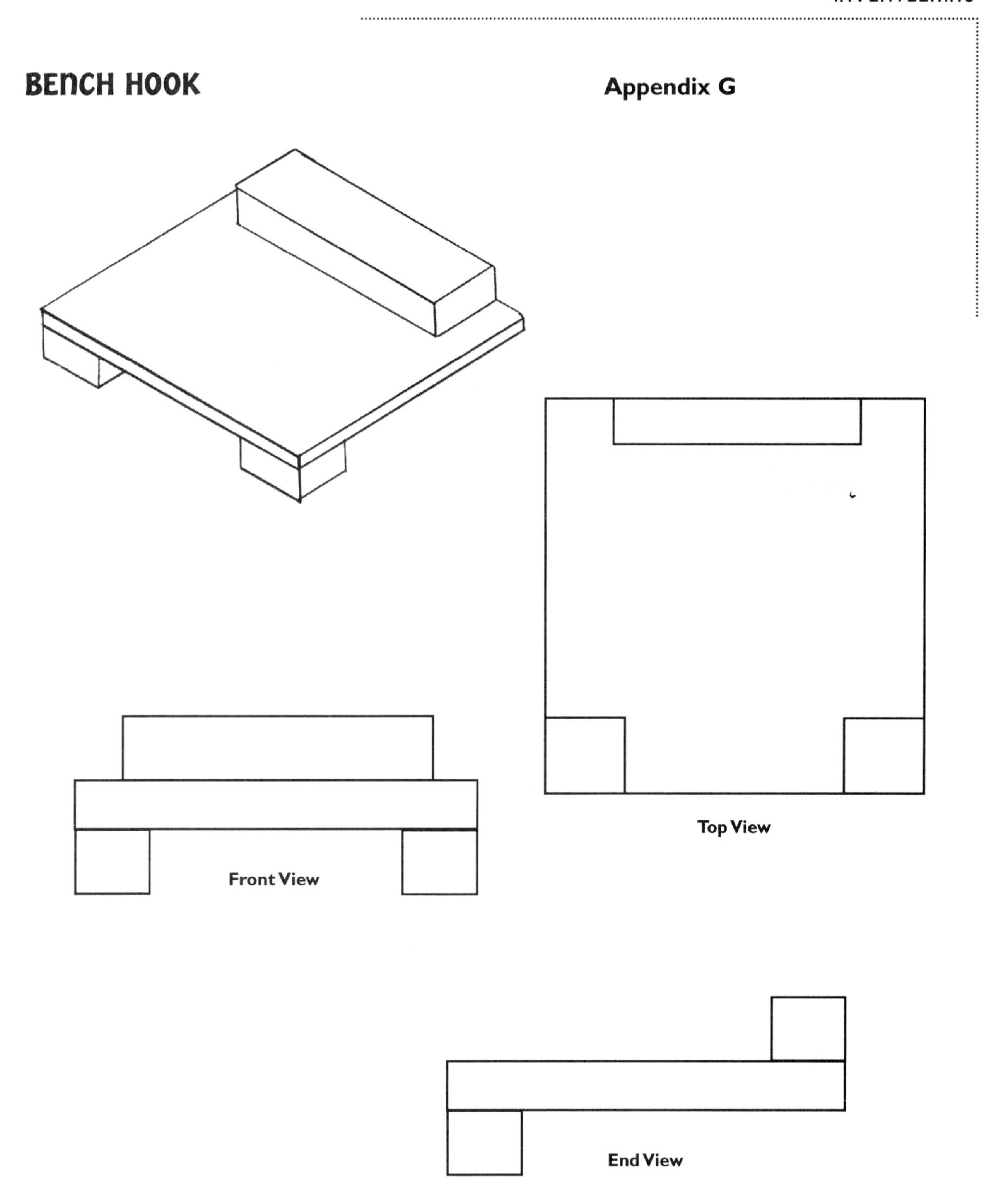

SOME INTERNET RESOURCES

Appendix H

Please note: The following websites were verified and bookmarked as of September 9th, 2000.

You can find out how a DVD works and explore other technology.
www.howstuffworks.com

Some different science and technology items, plus a parent's page.
www.hhmi.org/coolscience

This is a site from *Ranger Rick* magazine, and mostly covers nature and environment. Some games too.
www.nwf.org/kids

Mostly weather, environment, space.
explorezone.com

Many space-related activities and links
spacekids.hq.nasa.gov

IDEAS AND TECHNIQUES

The following resources are practical in nature and introduce teachers and children to the ideas, materials and techniques required to design and make things.

Caney, Steven; *Steven Caney's Invention Book,* Workman Publishing Co. Inc., 1985.ISBN 0-89480-076-0 — Explains the steps of the process of becoming an inventor. Has stories that tell the history of inventions we commonly know about. Also has a few start-up ideas or inspirations in a list of Fantasy Inventions that follow each story.

Catherall, Ed and Bev McKay; *Turning to Wheels,* Chrysalis Publications, 1988. ISBN 0-921049-13-7 — Reproducible activity pages of simple projects using wheels in more than 30 applications.

Chapman, C., et al. *Collins Technology for Key Stage 3: Design and Technology, the Process.* 1992, London: Collins Educational. (Good ideas to help you with the design process.)

Corney, Bob & Dale, Norman. *Technology I.D.E.A.S.* Canada. 1992. Maxwell MacMillan. (Projects for elementary students.)

Dalzell, Rosie; *We Can Join It,* Cherrytree Press Ltd., 1991. ISBN 0-7451-5128-0 — From the making of glue through paper folding, sewing, knotting, taping, and weaving, children learn how to use simple materials to make things.

Dalzell, Rosie; *We Can Move It,* Cherrytree Press Ltd., 1991. ISBN 0-7451-5129-9 — Centred around the theme of "moving things", children are introduced, through project applications, to rollers, trolleys, sliders, magnets, balloons, and pivots.

Gadd, Tim and Dianne Morton; *Technology Key Stage 1,* Stanley Thornes Ltd., 1992. Distributed in Canada through Bacon and Hughes. ISBN 0-7487-1357-3 — Huge bank of ideas, activities, and photo-copiable sheets. Demonstrates how technology can be introduced through themes and topics, stories and challenges. Also includes techniques and resources for record keeping and assessment.

Gadd, Tim and Dianne Morton; *Technology Key Stage 2,* Stanley Thornes Ltd., 1992. Distributed in Canada through Bacon and Hughes. ISBN 0-7487-1495-2 — Additional bank of ideas, activities, and photo-copiable sheets. Demonstrates how technology can be introduced through themes and topics, stories and challenges. Also includes techniques and resources for record keeping and assessment.

Harrison, Patricia and Chris Ryan, *Technology in Action: Unit 1,* Folens Ltd.,1990. ISBN 1-85276115-6 — A binder of activity sheets and teacher's resource booklet that introduces the topics of Ourselves,

The Farm, Homes and Toys. Each topic has several activities or challenges on laminated sheets. The supporting teacher resource also has an excellent section on using simple materials to develop basic skills.

Harrison, Patricia and Chris Ryan; *Technology in Action: Unit 2,* Folens Ltd., 1990. ISBN 1-85276116-4 — Same structure and resources as Technology in Action: Unit 1. The topics in this unit include Vehicles, Creatures, School, and The Park.

Horvatic, Anne; *Simple machines,* E. P. Dutton: New York; Fitzhenry and Whiteside, 1989. ISBN 0-525-44492-0 — A well illustrated picture book of simple machines used in places you might never expect. Everyday examples of the wheel, inclined plane, wedge, screw and lever are described.

Kerrod, Robin; *How Things Work,* Marshall Cavendish Corp., 1990. ISBN 1-85435-154-0 — Several well-illustrated and easy to follow pages of project ideas and how they work. Projects ranging from flight and simple propulsion to the making of electricity with a lemon, use simple and commonly available materials.

Knapp, Brian; *How Things Work,* Grolier Ltd., 1991. ISBN 0-7172-2783-9 From a tap works to a vacuum cleaner, this book plots a structural path of science, design, and technology.

Lampton, Christopher; *Seesaws, Nutcrackers and Brooms,* Millbrook Press, 1991. ISBN 1-878841-43-2 — An illustrated picture book describing simple machines that are really levers.

Malam, John; *Pop-up Machines,* Alfred A. Knopf Inc., 1991. ISBN 0-679-80872-8 — Five of today's strongest machines come to life off the pages. These pop-ups are contrasted with the way things used to be in the past.

Metropolitan Toronto School Board. *All Aboard! Cross-Curricular Design and Technology Strategies and Activities,* Toronto. 1996. Trifolium Books Inc. (Theme-based activities and teacher support, Grades K–6.)

Metropolitan Toronto School Board. *By Design, Technology Exploration & Integration,* Toronto. 1996. Trifolium Books Inc. (Problem-solving through design and technology. Activities and teacher support, Grades 6–9.)

Peel Board of Education Teachers *Mathematics, Science, & Technology Connections.* Toronto. 1996. Trifolium Books Inc. (Activities and planning assistance, Grades 6-9)

Reynolds, William, Corney, Bob, & Dale, Norman. *Imagineering – a "Yes, We Can!" Sourcebook for Early Technology Experiences.* Toronto. 1999 Trifolium Books Inc. (Activities and teacher support, Grades K–3)

Richards, Roy; *An Early Start To Technology,* Simon & Schuster Inc.,1990. ISBN 0-7501-0033-8 — A well illustrated and described gathering of experiences that make close links to technology and science. Provided is a wealth of practical experiences that will involve children in looking at structures, materials, forces, energy, and how things are controlled.

Richards, Roy. *An Early Start to Technology.* London. 1990. Simon & Schuster. (Emphasis on things children can readily design and make through problem-solving.)

Rockwell, Anne; *Machines,* Macmillan Publishing, 1972. ISBN 0-02-777520-8 — Illustrations and descriptions of everyday machines that make work easier.

Sellwood, Peter and Fred Ward and Ron Lewin; *Let's Make It Work,* MacMillan Education Ltd., 1990. ISBN 0-333-44023-4 This Introductory Book introduces children to simple technology and science through a problem-solving approach. Concepts and skills are systematically developed through simple work with different materials.

Stroud, Peter; *Tools,* Cherrytree Press Ltd., 1991. ISBN 0-7451-5148-5 A storybook introducing recognizable tools that help us do things that we cannot do with our bare hands.

Williams, John; *First Technology: Tools,* Wayland Publishers Ltd., 1993. ISBN 0-7502-0650-0 — A paper streamer activity enables children to use some basic tools that are clearly described and illustrated.

Williams, John; *Starting Technology: Machines,* Wayland Publishers Ltd.,1991. ISBN 0-7502-0025-1 — A book of creative ideas for making levers, diggers, cranes, pulleys and switches. The making of models support an understanding of how machines work.

Williams, John; *Starting Technology: Wheels,* Wayland Publishers Ltd., 1990. ISBN 0-7502-0271-8 — A book of creative ideas to make rollers, carts, trolleys, gears, and waterwheels to find out about movement.

Williams, Peter and Jacobson, Saryl, *Take a Technowalk to Learn about Materials and Structures.* ISBN 1-895579-76-7 — Provides teachers of Grades K–8 with 10 fun Technowalks designed to encourage students to investigate the materials and structures that surround us.

Williams, Peter and Jacobson, Saryl, *Take a Technowalk to Learn about Mechanisms and Energy* ISBN 1-55244-004-4 — Provides teachers of Grades K–8 with 10 fun Technowalks designed to introduce students to the concepts of mechanisms and energy by investigating the technologies all around us.

Zubrowski, Bernie; *Messing Around With Drinking Straw Construction,* Little, Brown and Co., 1981. ISBN 0-316-98873-1 — Directions for the making of drinking straw models to explore how houses, bridges and towers are made.

The resources listed in this section are grouped into themes commonly used in classrooms. These themes will be a source for developing ideas for additional challenges. In addition, they can be used to support the challenges withing this book.

COMMUNITIES

Bourgeois, Paulette and Kim LaFave; *Canadian Fire Fighters,* Kids Can Press, 1991. ISBN 1-55074-042-3 — Lively verses and illustrations tell the story of an hour in a small boy's life one hectic Monday morning.

Bourgeois, Paulette and Kim Lafave; *Canadian Garbage Collectors,* Kids Can Press, 1991. ISBN 1-55074-040-7 — An illustrated depiction of a day in the life of a garbage collector and highlighting the 3 Rs of Reduce, Reuse, and Recycle.

Bourgeois, Paulette and Kim LaFave; *Canadian Postal Workers,* Kids Can Press, 1992. ISBN 1-55074-058-X — This illustrated book traces the route of a letter written by a child to his Grandmother thousands of miles away. It also depicts a day in the life of a postal worker.

Cobb, Vicki; *Skyscraper Going Up!* Harper and Row Publishers Inc., 1987. ISBN 0-690-04525-5 — A pop-up book showing the "bones", "heart", "skin", and "lungs" of a building. Children can be the builders with bright, colourful paper mechanics guiding their participation.

Corcos, Lucille; *The City Book,* Western Publishing Co., 1972. ISBN 0-307-65772-8 — Watercolour illustrations and descriptions about cities and all the things that happen when people come together to build their homes, stores, and markets and to make and trade the things they make things they needed.

Gibbons, Gail; *Up Goes The Skyscraper!* Four Winds Press: Macmillan Publishing Co., 1986 ISBN 0-02-736780-0 — A clearly described and colourful construction of a skyscraper. Playing the part of people on the street, children can see it rise before their eyes.

Green, John F.; *Junk Pile Jennifer,* Scholastic Inc., 1991. ISBN 0-590-73680-9 A visual experience with Jennifer who loves junk. The neighbourhhood thinks she is crazy as she looks for new treasures to build a cosy house in the backyard.

Hoban, Tana; *Dig, Drill, Dump, Fill,* Greenwillow Books, 1975. ISBN 0-688-80016-5 — Photographs help children use their eyes and minds imaginatively with heavy machinery.

Jeunesse, Gallimard and Claude Delafosse; *On Wheels,* Moonlight Publishing Ltd., 1991. ISBN 1-85103-111-1 — A ride in a car, fire engine, racing car, ambulance, bulldozer, bicycle, and bus takes you on a trip through the sights and sounds of the city and countryside.

Rockwell, Harlow; *My Dentist,* Greenwillow Books, 1975. ISBN 0-688-80004-1 — Economic text combined with colourful illustrations of dentist office procedures and equipment.

Rockwell, Harlow; *My Doctor,* Macmillan Publishing Company Inc., 1973. ISBN 0-02-777480-5 — A picture book with short effective descriptions of the doctor's office and the medical equipment used.

Pfanner, Louise; *Louise Builds A House,* Orchard Books, 1989. ISBN 0-531-05769-8 — A pattern of simple words and absorbing pictures to satisfy any child who has ever longed to create a wonderful dream house. Louise's house appears page by page and piece by piece in imaginative detail.

Shefelman, Janice; *Victoria House,* Gulliver Books: Harcourt Brace Jovanovich, 1988. ISBN 0-15-200630-3 — A story with pictures of an old Victorian house moved from the country to its new location on a city street, where a family fixes it up and moves in.

Wilson, Forrest; *What It Feels Like To Be A Building,* The Preservation Press, 1988. ISBN 0-89133-147-6 — In simple and direct explanations and with lively drawings, children will find out about architecture. They will learn how they would feel if they were squashed, pushed, shoved, and tugged at, just as the parts of buildings are.

FAMILY & FRIENDS

Franklyn, Mary Eliza; *Pepper Makes Me Sneeze,* Petheric Press Ltd., 1978. ISBN 0-919380-25-5 — The junior chef is led through some interesting bits of Nova Scotia and with some helpful hints and reminders about what preparing food is all about.

Hughes, Shirley; *Moving Molly,* The Bodley Head Ltd., 1984. ISBN 0-370-30125-0 — A story about Molly and her family and the ordeals of moving.

Smith, Lucia B.; *My Mom Got A Job,* Holt, Rinehart and Winston, 1979. ISBN 0-03-048321-2 — Special times and days spent with mother are gone when mom goes to work. Some changes in life are fun and bring a new meaning to the whole family.

Stevenson, James; *The Sea View Hotel,* Greenwillow Books, 1978. ISBN 0-688-80168-4 — A skillfully paced read-aloud about a whopping tale concocted by Grandpa.

Spier, Peter; *Oh Were They Ever Happy,* Doubleday and Company Inc., 1978 ISBN 0-385-13175-5 — Resourceful children and one great idea, turn an unsupervised Saturday into a bedlam of paints, brushes, and pets underfoot.

Super, Gretchen; *What Kind Of Family Do You Have?* Twenty First Century Books, 1991. ISBN 0-941477-64-9 — A look at families like yours, unlike yours and families that are as different as the people who live in them.

PIONEERING

Adams, Peter; *Early Loggers And The Sawmill,* Crabtree Publishing Co., 1981. ISBN 0-86505-006-6 — An excellent set of early photographs and artist's illustrations with descriptions of early settlers and their logging and sawing operations.

Anno, Mitsumasa; *The Earth Is A Sundial,* The Bodley Head Ltd., 1984 ISBN 0-370-31016-0 — A pop-up book with activities designed to use shadows to tell time.

Gibbons, Gail; *Farming,* Holiday House Inc., 1988. ISBN 0-8234-0682-2 A picture book describing the activities and special qualities of life on the farm.

Gibbons, Gail; *The Milk Makers,* Macmillan Publishing Co., 1985. ISBN 0-02-736640-5 — Beginning with the cow, production, transportation, processing and final delivery of milk to the store is illustrated.

Goldreich, Gloria, Ester Goldreich and Robert Ipcar; *What Can She Be?— A Farmer,* Lothrop, Lee and Shepard Co., 1976. ISBN 0-688-41768-X The joys and satisfactions, as well as the hard work of running a farm, are conveyed in clear, simple language and photo-examples. Two farming sisters are followed through the changing seasons on a busy farm.

Greenwood, Barbara; *A Pioneer Story,* Kids Can Press Ltd., 1994.
ISBN 1-55074-237-X — A story of a pioneer family living on a backwoods farm in 1840. Following the family through the year until winter closes around them, children will learn what it is like to attend a backwoods school, weave cloth, and build a house.

Greenwood, Barbara; *Pioneer Crafts,* Kids Can Press Ltd., 1997.
ISBN 1-55074-359-7 — Illustrated step-by-step instructions to make crafts the same way pioneer children did. Crafts include making a rag doll, silhouette portrait, spatter stenciling, crazy quilt, moccasins, and candle making.

Humphrey, Henry; *The Farm,* Doubleday and Co., 1978.
ISBN 0-385-01388-4 — A photographic picture book helps children find out what a real farm is all about. Clear, simple text gives a realistic snapshot of contemporary farm life.

Kalman, Bobbie; *Early Schools,* Crabtree Publishing Co., 1982.
ISBN 0-86505-014-7 — Education, because of the work that had to be done, was for a time, a luxury the early settler could not afford. The book traces these early hardships, the school day and the consequences children faced.

Kalman, Bobbie; *Early Stores And Markets,* Crabtree Publishing Co., 1981.
ISBN 0-86505-004-X — A collection of photographs and artist's illustrations depicting life with the Trading Post and 17th and 18th Century store. Described are the many professions of the storekeeper, the goods sold, and the barter system.

Kalman, Bobbie; *Early Travel,* Crabtree Publishing Co., 1981.
ISBN 0-86505-008-2 — This book traces why early settlers had to travel, the means used, and the conditions they faced.

Kalman, Bobbie; *Early Village Life,* Crabtree Publishing Co., 1981.
ISBN 0-86505-010-4 — A collection of photographs and artist's illustrations traces an odyssey of life in the backwoods. Depicted are earliest examples of people sharing resources, working together, while also having a good time.

Kalman, Bobbie; *Food For The Settler,* Crabtree Publishing Co., 1982.
ISBN 0-86505-012-0 — Beginning with the first pioneers in the bush, the book traces the early growing of crops, raising animals, preparing dinners, and the making of bread and butter.

SPACE

Bantock, Nick; *Wings: A Pop-Up Book Of Things That Fly,* Random House, 1990. ISBN 0-679-81041-2 — Spectacular three-dimensional pop-ups and pull-tabs show children how wings really work on everything from bats and dragonflies to fighter planes and supersonic jets.

Catherall, Ed.; *Dropping In On Gravity,* Chrysalis Publications, 1988. ISBN 0-921049-14-5 — Pages of reproduceable activities that range from the making of a mobile to a game to launch a satellite, all support an understanding of the forces of gravity.

Cole, Joanna; *The Magic School Bus; Lost In The Solar System,* Scholastic Inc., 1990. ISBN 0-590-41429-1 — Blast off with Ms. Frizzle and her bus load of children on a field trip to the solar system.

Dixon, Malcolm; *Flight,* Wayland Publishers Ltd., 1990. ISBN 1-85210-931-9 From dandelion seeds and natures flight technology and hot-air balloons to rocket-propelled spacecraft, this book explains the technology behind many forms of flight. Activities are used throughout to support an understanding of these forms.

Francis, Neil; *Super Fliers,* Kids Can Press Ltd., 1988. ISBN 0-921103-37-9 A book beginning with discovering how flying fish and squirrels fly to the many mysteries of flight. Instructions are included to make and fly paper airplanes, helicopters, kites, parachutes, gliders, and more.

Little, Kate; *Things That Fly,* Usborne Publishing Ltd., 1987. ISBN 1-85123-204-4 — Colourfully illustrated explanations of how the first fliers to the worlds biggest and fastest planes stay up in the air.

Mackie, Dan; *Flight,* Hayes Publishing Ltd., 1986. ISBN 0-88625-112-5 Ideal for later primary grades in giving children an overlook of many sides of flight.

Mackie, Dan; *Space Tour,* Hayes Publishing Ltd., 1986. ISBN 0-88625-103-6 Ideal for later primary grades in giving children an opportunity to imagine they will be preparing for spaceflight.

Morris, Campbell; *Advanced Paper Aircraft Construction,* Angus and Robertson Publishers, 1989. ISBN 0-207-14502-4 — Easy to follow folding and throwing instructions for 14 darts, gliders and spinners.

Myring, Lynn; *Finding Out About Rockets And Spaceflight,* Usborne Publishing Ltd., 1982. ISBN 0-86020-584-3 — A picture information book that will answer many questions asked by children who are discovering the mysteries of space and space travel. Answers are given to such questions as; why do astronauts wear space suits?; can people live in space?; how do rockets work?; and what is a satellite?

Petty, Kate; *On A Plane,* Aladdin Books Ltd./Franklin Watts, 1984.
ISBN 0-531-04716-4 — Full colour art work and large easy-to-read text introduce children to what it is like to fly in a plane.

Williams, John; *Air,* Wayland Publishers Ltd., 1990.
ISBN 0-7502-0268-8 — While looking at air and indicators that it is there, and machines that use air to make them work, this book is full of fun ideas for making windmills, wind wheels, kites, and parachutes.

Williams, John; *Flight,* Wayland Publishers Ltd., 1991, ISBN 0-7502-0026-X
A book full of ideas for making paper darts, gliders, kites, helicopters and model birds. Notes are included on how this topic can be included in primary science and cross-curricular teaching.

TIME

Anno, Mitsumasa; *The Earth Is A Sundial,* The Bodley Head Ltd., 1984
ISBN 0-370-31016-0 — A pop-up book with activities designed to use shadows to tell time.

Gibbons, Gail; *The Reasons For Seasons,* Holiday House, 1995.
ISBN 0-8234-1174-5 — A pictorial explanation of how the position of the earth in relation to the sun causes our seasons and the wonders that come with them.

Humphrey, Henry and Deidre O'Meara Humphrey; *When Is Now?*
Doubleday and Company Ltd., 1980. ISBN 0-385-13215-8
A well-illustrated book to help children trace the development of time-keeping devices and to assist them in making their own models and to experiment with these devices.

Knapp, Brian; *Time,* Grolier Ltd., 1992. ISBN 0-7172-2875-4
A structure of diagrams, text, and activities to give children an idea how time influences everything from our body clock to the origin of the universe.

Kurth, Heinz; *Time,* The Windmill Press, 1973. ISBN 0-437-53605-X
Pictures and text look at some aspects of this mysterious medium called time. It shows how concepts of time have changed and progressed over the years.

Smith, A.G.; *What Time Is It?,* Stoddard Publishing Company Ltd., 1992.
ISBN 0-77375-525-X — A book filled with detailed illustrations of numerous timekeeping devices from every age; how calendars were invented; and how people kept track of time before clocks were used.

Webb, Angela; *Talk About Sand,* Franklin Watts Inc., 1992
ISBN 0-531-10370-6 — A well-illustrated book that will challenge children to suggest ideas for exploration and experimentation with sand. Activities include making a timer.

Yorke, Jane (Ed.); *Time,* Random House, 1991. ISBN 0-679-81164-8
A picture book with a first look at what happens in time from getting up in the morning to going to bed at night.

Zubrowski, Bernie; *Raceways: Having Fun With Balls And Tracks,* Greenwillow Books, 1985. Distributed by Gage Educational Publishing Co. ISBN 0-688-04159-0 — Using plastic decorative molding available in lumber supply stores, children can make marbles go up and down hills, travel in a circle, and jump from track to track without falling off. Travel times as well as basic scientific ideas such as energy, acceleration, and momentum can be linked to a variety of illustrated games and experiments to try.

RECOMMENDED CONSTRUCTION KITS

Duplo. (Larger version of Lego designed for young children. Excellent introduction to Lego. Permits imaginative play.)

Lasy (Imaginit). (Plastic rectangles, connectors, etc.)

Stickle Bricks. (Squares, triangles, and other pieces that "stick" together. Can be combined with "straws." Easy to use.)

Baufix. (Wood and plastic, similar to Meccano. Students use nuts, bolts, and spanners. Durable and colourful.)

Lego. (Very versatile. Suggest transferring the pieces into a large box with many small compartments, rather than the original cardboard box. Some components are fragile.)

Meccano. (Original version can be difficult for younger students, but there is a new set with plastic that is easier for them to use.)

Ramagon. (Used by NASA engineers to model space stations. Easy to use and works with other construction kits. Durable.)

Other Science and Technology Titles from Trifolium Books

SPRINGBOARDS FOR TEACHING SERIES

INVENTEERING
A Problem-Solving Approach to Teaching Technology

Bob Corney & Norm Dale

An essential "getting started" resource for teachers of **Grades K–8** wanting to provide their students with hands-on technological experiences.

8½" × 11" • 128 PAGES • SOFT COVER
ILLUSTRATIONS • ISBN: 1-55244-014-1
$29.95 CAN. • AVAILABLE 2000 © 2001

IMAGINEERING
A "Yes, We Can!" Sourcebook for Early Technology Experiences

Bill Reynolds, Bob Corney, and Norm Dale

Packed with ideas to stimulate young students' imagination and creativity as they explore the issues and applications of technology. For teachers of **Grades K–3.**

8½" × 11" • 144 PAGES • SOFT COVER
ILLUSTRATIONS • ISBN: 1-895579-19-8
$29.95 CAN. • AVAILABLE

ALL ABOARD!
Cross Curricular Design and Technology Strategies and Activities

By Metropolitan Toronto School Board teachers

This teacher-tested resource helps educators integrate design and technology easily and effectively into day-to-day lessons. For teachers of **Grades K–6.**

8½" × 11" • 176 PAGES • SOFT COVER
ILLUSTRATIONS • ISBN: 1-895579-86-4
$21.95 CAN. • AVAILABLE

Take a Technowalk to Learn about Materials and Structures

Peter Williams & Saryl Jacobson

Provides teachers of **Grades K–8** with 10 fun Technowalks designed to encourage students to investigate the materials and structures that surround us.

8½" × 11" • 96 PAGES • SOFT COVER
ILLUSTRATIONS • ISBN: 1-895579-76-7
$21.95 CAN. • AVAILABLE

Take a Technowalk to Learn about Mechanisms and Energy

Peter Williams & Saryl Jacobson

Provides teachers of Grades K–8 with 10 fun Technowalks designed to introduce students to the concepts of mechanisms and energy by investigating the technologies all around us.

8½" × 11" • 92 PAGES • SOFT COVER
ILLUSTRATIONS • ISBN: 1-55244-004-4
$25.95 CAN. AVAILABLE

NEW FOR 2000 — *"Experimenting with..." series*

Fun and interesting activities introduce Grades 4–8 students to hands-on science. Students will develop analytical skills and creative thinking while learning about their physical world.

Experimenting with Air

By Gordon R. Gore

8½" × 11" • 46 PAGES • SOFT COVER
ILLUSTRATIONS • ISBN: 1-55244-042-7
$10.00 CAN. • AVAILABLE

Experimenting with Electricity

By Gordon R. Gore

8½" × 11" • 46 PAGES • SOFT COVER
ILLUSTRATIONS • ISBN: 1-55244-040-0
$10.00 CAN. • AVAILABLE

Experimenting with Forces

By Gordon R. Gore

8½" × 11" • 46 PAGES • SOFT COVER
ILLUSTRATIONS • ISBN: 1-55244-032-X
$10.00 CAN. • AVAILABLE

Experimenting with Energy

By Gordon R. Gore

8½" × 11" • 46 PAGES • SOFT COVER
ILLUSTRATIONS • ISBN: 1-55244-044-3
$10.00 CAN. • AVAILABLE

Experimenting with Simple Machines

By Gordon R. Gore

8½" × 11" • 46 PAGES • SOFT COVER
ILLUSTRATIONS • ISBN: 1-55244-038-9
$10.00 CAN. • AVAILABLE

Experimenting with Light and Colour

By Gordon R. Gore

8½" × 11" • 46 PAGES • SOFT COVER
ILLUSTRATIONS • ISBN: 1-55244-036-2
$10.00 CAN. • AVAILABLE

Experimenting with Physical and Chemical Changes

By Gordon R. Gore

8½" × 11" • 46 PAGES • SOFT COVER
ILLUSTRATIONS • ISBN: 1-55244-034-6
$10.00 CAN. • AVAILABLE

TEACHERS HELPING TEACHERS SERIES

BY DESIGN
Technology Exploration and Integration

By the Metropolitan Toronto School Board Teachers

Over 40 open-ended activities for **Grades 6–9** integrate technology with other subject areas.

8½" × 11" • 176 PAGES • SOFT COVER
ILLUSTRATIONS • ISBN: 1-895579-78-3
$39.95 CAN. • AVAILABLE

Mathematics, Science, & Technology Connections

By Peel Board of Education Teachers

Twenty-four exciting integrated Math, Science, and Technology activities for **Grades 6–9**.

8½" × 11" • 160 PAGES • SOFT COVER
ILLUSTRATIONS • ISBN: 1-895579-37-6
$39.95 CAN. • AVAILABLE